# Soviet Defense Spending

TEXAS A&M UNIVERSITY MILITARY HISTORY SERIES

58

# Soviet Defense Spending

## A History of CIA Estimates, 1950–1990

*Noel E. Firth
and
James H. Noren*

Texas A&M University Press
College Station

Copyright © 1998 by Noel E. Firth and James H. Noren
Manufactured in the United States of America
All rights reserved
First edition

This manuscript was originally prepared under the sponsorship
of the Central Intelligence Agency, Contract No. 96.N.349309.000.

The views expressed in *Soviet Defense Spending: A History of CIA Estimates,
1950–1990* are not necessarily those of the Central Intelligence Agency.

The paper used in this book meets the minimum requirements
of the American National Standard for Permanence
of Paper for Printed Library Materials, Z39.48-1984.
Binding materials have been chosen for durability.

Library of Congress Cataloging-in-Publication Data

Firth, Noel E., 1933–
    Soviet defense spending : a history of CIA estimates., 1950–1990 /
Noel E. Firth and James H. Noren.
        p.  cm. — (Texas A&M University military history series ; 58)
    Includes bibliographical references (p.   ) and index.
    ISBN 0-89096-805-5
    1. Soviet Union—Armed Forces—Appropriations and expenditures.
2. Cold War. 3. Intelligence service—United States. I. Noren, James H.,
1929– . II. Title. III. Series.
UA770.F467   1998
355.6´2247—dc21                                             97-47002
                                                                        CIP

*We dedicate this book
to the many men and women
who worked throughout the Cold War
with integrity and professionalism
to create and continually improve
the important analytic capability
of the U.S. intelligence system
that is described in this book.*

# Contents

| | |
|---|---|
| List of Figures | ix |
| List of Tables | xi |
| Preface | xiii |
| Acknowledgments | xv |
| List of Abbreviations | xvii |

CHAPTERS

| | | |
|---|---|---|
| 1. | Introduction | 3 |
| 2. | Getting Started, 1950–60 | 9 |
| 3. | New Questions, New Techniques, and New Evidence, 1961–74 | 34 |
| 4. | The Estimates, 1975–90: Substantial Progress and Managerial Doubts | 57 |
| 5. | The Spending Estimates | 98 |
| 6. | The Critics | 140 |
| 7. | The Defense Spending Estimates in Perspective | 192 |

| | |
|---|---|
| Appendix A. Costing Improvements, 1975–90 | 209 |
| Appendix B. The Index Number Problem and Defense Programs: Some Considerations | 223 |
| Notes | 227 |
| References | 265 |
| Index | 283 |

# Figures

| | | |
|---|---|---|
| 4.1. | Key Events in Military-Economic Analysis | 58 |
| 4.2. | Soviet Defense Outlays | 82 |
| 4.3. | Soviet Defense Procurement | 85 |
| 5.1. | Soviet Defense Outlays | 100 |
| 5.2. | Soviet Defense Spending | 102 |
| 5.3. | USSR: Average Annual Spending by Resource Category | 106 |
| 5.4. | Distribution of Defense Spending by Resource Category | 107 |
| 5.5. | USSR: Average Annual Spending by Mission | 111 |
| 5.6. | Distribution of Defense Spending by Mission | 112 |
| 5.7. | Dollar Cost of U.S. and Soviet Military Programs, 1951–64 | 113 |
| 5.8. | Dollar Cost of U.S. and Soviet Military Programs, 1965–89 | 116 |
| 5.9. | Dollar Cost of Soviet and U.S. Military Programs by Mission | 117 |
| 5.10. | Dollar Cost of Soviet and U.S. Military Programs by Resource Category | 117 |
| 5.11. | Defense as a Share of Soviet GNP under Alternative Definitions of National Security | 134 |
| 6.1. | Ruble and Dollar Comparisons of Soviet and U.S. Defense Activities | 149 |
| 6.2. | Soviet Procurement in Two Types of Constant Rubles | 156 |

| | | |
|---|---|---|
| 6.3. | Effect of Expansion of the Price Sample on Procurement Estimate | 166 |
| 6.4. | Reduction in Absolute Error Using Aggregated Estimates | 174 |
| 6.5. | Estimates of Soviet Defense Spending Based on the Official Defense Budget | 176 |
| 6.6. | Residual Methodology | 180 |
| 6.7. | Estimates of Soviet Military Procurement | 181 |
| 6.8a. | Comparison of Lee's and CIA's Residual Estimates of Soviet Military Machinery Purchases | 182 |
| 6.8b. | Lee's Residual and Lee's Residual Adjusted by CIA | 182 |
| 6.9. | Soviet Defense Expenditures | 189 |
| A.1. | Best Estimate of Soviet Military Manpower | 212 |
| A.2. | Process of Estimating the Resource Costs of Soviet Military RDT&E | 220 |
| A.3. | Estimated Soviet Military RDT&E Expenditures | 220 |
| A.4. | Soviet Total RDT&E Expenditures | 221 |

# Tables

| | | |
|---|---|---|
| 2.1. | Spending Estimates by Economic Resource Category | 15 |
| 2.2. | Spending Estimates by Military Service | 16 |
| 2.3. | Spending Estimates by Military Mission | 17 |
| 3.1. | Soviet Defense Spending in 1968 | 53 |
| 4.1. | CIA's Estimates of Cumulative Defense Spending, 1971–75 | 60 |
| 4.2. | Emigré's Reconstruction of "Expenditures of Defense" Section of the Otsenka | 63 |
| 4.3. | Dates of Latest Dollar Cost Studies | 67 |
| 4.4. | Record of Published Estimates of Soviet Defense Spending | 76 |
| 5.1. | USSR Real Defense Spending by Resource Category | 101 |
| 5.2. | USSR Distribution of Cumulative Defense Spending by Resource Category | 106 |
| 5.3. | USSR Real Defense Spending by Mission | 108 |
| 5.4. | USSR Distribution of Cumulative Defense Spending by Mission | 111 |
| 5.5. | Comparisons of U.S. Defense Outlays and Dollar Valuations of Soviet Military Programs | 114 |
| 5.6. | U.S. and USSR: Distribution of Defense Spending in Terms of Dollars | 118 |
| 5.7. | Value of Non-Soviet Warsaw Pact and Non-U.S. NATO Defense Activities | 119 |

| | | |
|---|---|---|
| 5.8. | Dollar Value of Warsaw Pact and NATO Defense Activities by Resource Category | 120 |
| 5.9. | Landmarks in the Soviet Military Buildup | 122 |
| 5.10. | Ratio of Soviet Defense Spending to GNP | 129 |
| 5.11. | Estimated Costs of Selected U.S. and Soviet National Security Activities | 133 |
| 6.1. | Comparison of Soviet Defense Outlays | 151 |
| 6.2. | Cost Ratios across Soviet Weapon Generations | 161 |
| 6.3. | 1982 Ruble to 1988 Dollar Ratios Used in a CIA Comparison of U.S. and Soviet GNP | 168 |
| 6.4. | CIA's Historical Estimates of the Dollar Value of Soviet Defense Activities | 171 |
| 6.5. | Estimates of Soviet Defense Spending and Confidence Intervals for 1960, 1970, and 1980 | 174 |
| 6.6. | Published Structure of Soviet National Income by End Use | 178 |
| 6.7. | Alternative Estimates of Growth of Soviet Machine-Building Output | 184 |
| 6.8. | Official Defense Budgets for 1989 and 1990 | 186 |
| 6.9. | Reconstruction of Soviet Defense Budgets | 187 |
| A.1. | Estimates of Soviet Military Manpower for Mid-1988 | 211 |
| A.2. | Changes in Estimates of Ruble Outlays for Operations and Maintenance | 214 |

# Preface

This book is intended for those interested in the history of the Cold War, the history of U.S. intelligence, and studies of the former Soviet Union. As one former senior officer of the Central Intelligence Agency (CIA) pointed out to us, the CIA's estimates of Soviet defense spending for most of its history were in the middle of one of the biggest and most enduring political battles in Washington: the annual struggle over the size of the U.S. defense budget and how it would be allocated among competing programs.

Although CIA estimates of Soviet defense spending were often at the center of governmental and public debate throughout the Cold War, only summary information about them could be released to the public domain. The release of greater detail would have required the revelation of sensitive intelligence sources and methods. Most of this sensitivity has vanished now, and we can take a closer look at CIA estimates and how and why they were done. As such, our goals are modest. For example, CIA estimates often had an important impact on U.S. defense policy and programs, and we duly note this. But that impact is not the focus of the story, nor do we try to measure it with a high degree of precision; we gladly leave this task to others. We believe the story of the development of a significant intelligence analytic capability is an interesting one, and we hope readers will agree.

Many historians maintain that participants make poor historians of events, for as participants they are too involved to achieve the detachment necessary to write objectively. The other side of the argument is, of course, that participants in a process are far better able to understand the nuances and subtleties that never reach the written page, and to use that understanding to interpret the written record. We were both involved in much of what is presented in this study. James H. Noren spent nearly all of his thirty-two-year government career at the CIA conducting research and analysis on the Soviet economy. Noel E. Firth spent the first eighteen years of his thirty-one-year CIA career working directly on the development of estimates of Soviet defense spending. To the extent that this book reflects our recollections, it is properly viewed as personal memoir; but to the extent that it reflects the broad search of the written record—we reviewed thousands of pages of documents—and many interviews with producers and users of CIA's estimates, it is properly viewed as history. (The interviewees included a former president and other senior policymakers, as well as several former directors of Central Intelligence and other senior intelligence managers.) The net result is something between memoir and history. We trust that understanding the potential traps helped us avoid them. In any event, we are confident that many will let us know where we have failed.

This book was sponsored by the Center for the Study of Intelligence of the Central Intelligence Agency. The center provided outstanding production support to get the manuscript into finished form and provided access to the necessary source materials. The product of our efforts, however, should not be viewed as an officially approved history. We conducted research and analysis with complete independence. Official review of the manuscript was limited to ensuring that we had fulfilled our contractual obligation to produce a retrospective monograph on CIA's analysis of Soviet defense spending. Official guidance about specific contents and tone was neither sought nor offered. The views and judgments expressed are ours.

# Acknowledgments

We are indebted to the following individuals who read our draft manuscript and made important contributions to the final product through their comments and suggestions: Bob Abbott, Abraham Becker, Don Burton, Bruce Clarke, Terry Dunn, Harry Gelman, Grey Hodnett, Henry Nowik, Edward Proctor, John Reynolds, Gertrude Schroeder Greenslade, Derk Swain, and Doug Whitehouse. We, of course, accept full responsibility for any errors that may remain. We also thank Joyce Desmond and Kelly Petrilli, who typed more versions of the manuscript than any of us care to remember; Alice Bailey, whose computer and statistical support was indispensable; the Multimedia Production Group of the DCI Production Center for its patience and expertise in preparing our many and complex graphics; and Charles Dolgas whose assistance was both generous and invaluable. Finally, we thank our wives, Joan Firth and Alice Noren, who endured much during the preparation of the book.

# Abbreviations

| | |
|---|---|
| ABM | Anti-ballistic missile |
| ACDA | Arms Control and Disarmament Agency |
| ADP | Automatic data processing |
| AEC | Atomic Energy Commission |
| APC | Armored personnel carrier |
| BAM | Baikal-Amur Railroad (new trans-Siberian railroad) |
| CEMA | Council for Economic Mutual Assistance |
| CER | Cost-estimating relationship |
| CFE | Conventional Forces Europe |
| CIA | Central Intelligence Agency |
| CIG | Central Intelligence Group (predecessor organization to CIA) |
| CIS | Commonwealth of Independent States |
| CNO | Chief of Naval Operations |
| CPSU | Communist Party of the Soviet Union |
| DCI | Director of Central Intelligence |
| DCID | Director of Central Intelligence Directive |
| DDCI | Deputy director of Central Intelligence |
| DDI | Deputy director for Intelligence, CIA |
| DDR&E | Deputy for Defense Research and Engineering, Department of Defense |
| DI | Directorate of Intelligence, CIA |
| DIA | Defense Intelligence Agency, Department of Defense |
| DoD | Department of Defense |

| | | |
|---|---|---|
| FRG | Federal Republic of Germany |
| GDR | German Democratic Republic |
| GNP | Gross national product |
| GPO | U.S. Government Printing Office |
| HPSCI | House Permanent Select Committee on Intelligence, U.S. Congress |
| IAC | Intelligence Advisory Committee |
| IAP | Intelligence Assumptions for Planning |
| ICBM | Intercontinental ballistic missile |
| INF | Intermediate nuclear forces |
| IRBM | Intermediate-range ballistic missile |
| IISS | International Institute for Strategic Studies |
| JEC | Joint Economic Committee, U.S. Congress |
| KGB | Komitet gosudarstvennoy bezopasnosti (Committee for State Security), USSR |
| MBMW | Machine-building and metal-working |
| MEAP | Military-Economic Advisory Panel |
| MEBORR | CIA's first computerized system for estimating Soviet defense spending, named for the Military Expenditures Branch, Office of Research and Reports |
| MIC | Military-industrial complex |
| MOD | Ministry of Defense, USSR |
| MRBM | Medium-range ballistic missile |
| MVD | Ministerstvo vnutrennikh del (Ministry of Internal Affairs), USSR |
| NATO | North Atlantic Treaty Organization |
| NFAC | National Foreign Assessment Center, CIA |
| NIC | National Intelligence Council |
| NIE | National Intelligence Estimate |
| NIO | National Intelligence Officer |
| NIO/SP | National Intelligence Officer for Strategic Programs |
| NSA | National Security Agency |
| NSC | National Security Council |
| NSCID | National Security Council Intelligence Directive |
| NSWP | Non-Soviet Warsaw Pact |
| O&M | Operations and maintenance |

| | | |
|---|---|---|
| OCS | Office of Computer Services, CIA | |
| OER | Office of Economic Research, CIA | |
| OMB | Office of Management and Budget, Office of the President | |
| ONE | Office of National Estimates, CIA | |
| ORE | Office of Research and Estimates, CIA | |
| ORR | Office of Research and Reports, CIA | |
| OSD | Office of the Secretary of Defense | |
| OSI | Office of Scientific Intelligence, CIA | |
| OSR | Office of Strategic Research, CIA | |
| OSS | Office of Strategic Services (predecessor organization to CIA) | |
| PA&E | Program Analysis and Evaluation, Department of Defense | |
| PFIAB | President's Foreign Intelligence Advisory Board | |
| POL | Petroleum, oil, and lubricants | |
| R&D | Research and development | |
| RDT&E | Research, development, test, and evaluation | |
| SAC | Strategic Air Command | |
| SACEUR | Supreme Allied Command Europe | |
| SALT | Strategic Arms Limitation Talks | |
| SAM | Surface-to-air missile | |
| SCAM | Strategic Cost Analysis Model | |
| SDI | Strategic Defense Initiative | |
| SIPRI | Stockholm International Peace Research Institute | |
| SLBM | Submarine-launched ballistic missile | |
| SOVA | Office of Soviet Analysis, CIA | |
| SOVSIM | CIA econometric simulation of the Soviet economy | |
| SSCI | Senate Select Committee on Intelligence, U.S. Congress | |
| START | Strategic Arms Reduction Talks | |
| USIB | United States Intelligence Board | |
| WP | Warsaw Pact | |

Abbreviations

# Soviet Defense Spending

# Chapter 1
# Introduction

This book tells the story of the Central Intelligence Agency's (CIA) more than forty-year effort to estimate Soviet defense spending. It examines why and how the work was done, results of the estimating effort, and how the estimates were used. The book also addresses some of the criticisms of CIA estimates and what we have learned, since the collapse of the Soviet Union, about Soviet military outlays during the Cold War. We recognize that the story is in some respects incomplete. For instance, we rarely mention personalities and include only those bureaucratic events and decisions essential to an understanding of the estimates' evolution. The spending estimates themselves are the true focus of the story. Arguably, they were some of the most important numbers produced by CIA analysts during the Cold War.

First things first. The fundamental question raised about CIA's work in this area is: "Why bother with expenditure estimates at all? If you want to know something about Soviet military programs, wouldn't it be better to study the physical attributes of the defense establishment, unobscured by what has been called in the literature of economic theory 'the veil of money'?" The answer is yes and no. It is true that a direct examination of the ships, tanks, missiles, and aircraft that make up the order-of-battle or that are in production provides more vivid images of individual defense programs than images conveyed by monetary values. But it is also true that the mind has no way to assimilate diverse physical images into analytically meaningful constructs without the aid of some common denominator. It simply cannot be done.[1]

Money, performing its function as a unit of account, provides just such a common denominator. Monetary values make it possible to comprehend the magnitudes and trends in military programs in terms that take into account both the quantitative and qualitative dimensions of the underlying physical base. It takes no more than a moment's reflection on this particular function of money to realize what a truly amazing achievement and analytic benefit this ability to measure is. Analysts can apply a standardized measure of the resources committed to any specifically defined activity and gain an appreciation of the absolute and relative sizes of that activity. Through this standardized measure of resource commitment, which we generally refer to as a monetary valuation, we can easily compare unlike things, such as army divisions with naval aircraft carriers, or country A's defense program with country B's, or country A's program this year with that of last year. Money, however, is far from being a perfect, unambiguous measure of defense or any other human activity. Conceptual ambiguities such as those posed by the "index number problem" serve as constant reminders that monetary values of any activity are best viewed as approximations of reality, not perfect reflections.[2] Nevertheless, like democracy in the political arena, monetary measures are far better suited to the purpose at hand than anything else that has come along.

In the process of combining the vast amounts of physical data involved in military programs, the use of money as a common denominator overcomes an important flaw in human perception that afflicts intelligence analysts along with everyone else. People share a pronounced tendency to focus on things that are occurring and to ignore those that are not. For instance, suppose the production of a new military aircraft begins in year $t$ and the production of an older aircraft ends in year $t$ *minus one*. Hardly any notice will be taken, even by the analysts directly involved, of the fact that the older system is no longer being produced in year $t$. The production of the new system will be uppermost in analysts' minds and is likely to feature prominently in intelligence publications and thereby in the minds of intelligence consumers in the national security community. Both events, however—the new production and the cessation of old production—will register equally with the military-economic analyst charged with putting a monetary value on total aircraft production.[3]

Although the monetary estimates could and did make an important contribution to understanding the dimensions of the Soviet military effort, they were also sometimes employed—particularly in the political arena—to make unwarranted judgments about the military effectiveness of the Soviet effort. CIA estimates, by their nature, measured the monetary costs of inputs—manpower, materials, and other economic resources—to the military establishment. They did not attempt to measure the military effectiveness—that is, the output—produced by the inputs. The basic problem is that, unlike the inputs, which are objective, identifiable, and measurable, military effectiveness is subjective and usually difficult, if not impossible, to measure. For instance, to a fighter pilot military effectiveness may mean something like his ability in conjunction with his aircraft to destroy enemy aircraft. To a U.S. president it may mean the ability of the military establishment to deter hostile military, political, or economic actions by potential adversaries.

At times in the political debate over the U.S. defense budget, participants appeared to make the implicit assumption that expenditures for inputs are a perfect proxy for—that is, they are directly proportional to—military effectiveness. From a commonsense perspective, it would seem reasonable to assume that a generally positive correlation exists between the inputs to a military program and its effectiveness, particularly when the inputs are considered cumulatively over a period of years. But even this more relaxed assumption could be egregiously misleading. Those who view military effectiveness in terms of deterrence, for example, may worry that the commitment of resources to military programs could upset a precarious balance and end up producing less rather than more deterrence. Thus the relationship between inputs and outputs would not even be direct. The bottom line is that although the expenditure estimates conveyed unique and significant information about the Soviet defense effort, judgments about military effectiveness generally require a broader frame of reference and substantially more information than that conveyed by estimates alone.

The CIA was established officially on September 18, 1947. A little more than a week later, on September 26, the agency issued a document identified as CIA 1—its first major analytic publication. The CIA 1's weighty title, *Review of the World Situation as it Relates to the Security of the United States,* reveals much about the scope of the new agency's concerns and the vital nature of its responsibilities. A reading of the document itself provides important insights into the U.S. national security community's thinking in the early years of the Cold War. What comes through loud and clear is the recognition of the link between the economic-industrial strength of the USSR and its military power, and the centrality of this link to the achievement of Soviet political objectives, which were seen as unequivocally malevolent.

A major lesson of World War II had been learned: industrial potential is the key to military-political victory. It is not surprising that assessing the capabilities and potential of the Soviet economy became a high-priority analytic objective of the newly established intelligence community. After some initial discussion and debate, CIA was tasked by the National Security Council (NSC) with the primary responsibility for such analysis. The first questions for the national security policy community concerned the size of the Soviet defense budget and the impact of defense spending on the economy. These two questions about the Soviet defense effort—size and burden—remained at the core of CIA's military-economic analytic effort until the collapse of the Soviet Union.

These military-economic questions posed a formidable challenge to the fledgling CIA. Secrecy was the watchword of the Soviet approach to military data and indirectly related economic data. Even general information about the postwar Soviet economy was scarce, and much of what was available was of dubious reliability. Official information about defense spending was exceedingly sparse.[4] Over the next few years staff were recruited, methodologies developed, and data collected to build an independent capability for understanding the economic dimensions of Soviet defense activity.

From the beginning, CIA spending estimates played both analytic and advocacy roles in the national security process. At the analytic level, defense planners, arms control analysts, congressional staffers, Office of Management and Budget (OMB) analysts, NSC staff officers, and others made extensive use of the CIA product in their own work. Although the focus ranged over time from macro issues like levels and trends in total defense spending and defense shares of gross national product (GNP) to the specific micro issues that prevailed during the MacNamara era of cost-benefit analyses—for example, the five-year operating costs of a Yankee class submarine—the overall demand for CIA's military-economic analyses of the USSR remained high throughout the Cold War. Indeed, the continuing pressure to satisfy demand for military spending analyses eventually led to a serious error in CIA work.

At the advocacy level, policy officers and politicians discovered that the spending estimates could be an effective tool for achieving institutional ends in ongoing battles for programs and budgets. Not surprisingly, CIA producers of the spending estimates were accused by those who were not pleased with the results of being incompetent or politicizing their analyses to support a particular policy or their own ideological orientations. Many of the allegations came from defense "hawks" who saw low CIA estimates as undermining support for defense programs they favored. Other equally vehement allegations came from defense "doves," some of whom believed

CIA analysts were in league with the Pentagon to exaggerate Soviet defense spending in order to support higher U.S. defense budgets.

The highly charged atmosphere surrounding the spending estimates throughout most of their history raises the question of the degree to which they were, in fact, politicized by CIA. Politicization is a complex phenomenon that can take many forms. At one end of the spectrum is overt direction from above to "cook the books," and at the other end is the subtle but clear message from above that certain analytic conclusions are unwelcome and would be better left unpursued or uncommunicated. Our personal experiences and the research for this study confirm that the books were professionally honest and never cooked for political purposes. Questions, however, can be raised about the effect of some of the more subtle forms of politicization on the estimating process at certain points in time. These questions are addressed in chapter 4, and conclusions on the subject are presented in chapter 7.

In chapters 2, 3, and 4 we trace the evolution of military spending analyses through three chronological periods:

1. *1950–60, when the foundation for the work was laid, methodologies were explored and adopted, and military-economic analysis became an integral element of key national intelligence estimates (NIEs) dealing with Soviet capabilities and intentions.*
2. *1961–74, a period of dramatic growth in the demand for spending analyses and rapid improvement in costing techniques, but also a period when, at least until about 1972, a crucial element of the database—Soviet prices for defense hardware—was generally neglected.*
3. *1975–90, a period marked at its beginning by a major upward revision in ruble estimates resulting from a combination of belated research on Soviet military prices begun in 1972 and important new intelligence information. The sudden, large upward revision severely undermined confidence in CIA's costing work and stimulated a period of intensive research. This major research effort produced substantial improvement in the quality of databases underlying the spending estimates, as well as refinement and improved documentation of costing methodologies. Ironically, it was also a period when the support of costing work by senior CIA managers eroded.*

Chapter 5 presents the CIA estimate of Soviet defense spending as it had evolved by the late 1980s. Levels and trends in spending on military mission and resource category bases are laid out in detail. The many-faceted issue of the economic burden of Soviet defense activity is discussed, and comparisons are made with U.S. defense spending. Finally, the estimate is reviewed in the Soviet domestic and international political context in which the defense spending occurred; this serves as a check on the consistency of the estimate with key historical events of the period.

No story of CIA's analytic work on Soviet defense spending would be complete without consideration of the many criticisms it generated within and outside the government. The role of the spending estimates in the U.S. internal political debate and the secrecy surrounding the details of CIA's methodologies and databases combined to make attacks on CIA's work inevitable. The nature of the attacks runs the gamut of sophistication from intuitive rejection to well thought out, constructive questioning of concepts, measures, and data. Chapter 6 discusses the most important of these criticisms; individual critics who have raised essentially the same objection to CIA analyses are lumped together for purposes of a single discussion. The issues discussed in this chapter are often arcane and the argumentation is complex, but we believe that readers are entitled to sift through enough detail to make an independent assessment of the critics' views.

Chapter 7 presents some summary judgments and evaluations of the CIA military-economic analytic effort. How good was it? What contributions did it make to the U.S. intelligence process and U.S. national security policy? And the fundamental question: Were the benefits, if any, worth the cost or would U.S. taxpayers have been better served if the work had never been done?

We try to tell the story in a manner that will maintain the generalist's interest and provide enough technical detail to satisfy the specialist. Most of the more technical discussion appears in chapters 4 and 6 and the appendix.

# Chapter 2
# Getting Started, 1950-60

## Concepts and Methods

The first day at a new school, the first page of writing a book, the first five minutes of a blind date, the first year for a new business—there is no question about it, getting started is often the most difficult phase of human and institutional endeavors. Building a competent analytic capability for understanding the nature and dimensions of the Soviet military-economic threat to the West was no exception to this rule.

The goal was clear enough. The central national security policy issue in the early days of the Cold War was the need for a comprehensive understanding of Soviet military capabilities and intentions. The derivative economic question was whether or not the Soviet economy was capable of sustaining a military effort large enough to be a significant threat. While the ultimate goal was clear, the road to it was not. First, substantial bureaucratic potholes impeded progress and required adroit maneuvering. And then there was the daunting challenge of developing methodologies to see through the dense fog of secrecy and unfamiliarity that obscured the Soviet economy. Western experts had no experience analyzing the workings of a closed, nonmarket economy.

### The CIA Charter for Analysis of Soviet Defense Spending

The best term to describe CIA's authority for analysis of the Soviet economy when the agency was established is "squatter's rights." There was only vague written authority at the outset,[1] but to the extent that the capability to perform such analysis existed in the U.S. government, it could be found in CIA in the form of the staff inherited from the Central Intelligence Group (CIG), the follow-on organization to the wartime Office of Strategic Services (OSS). The CIG's Office of Research and Estimates (ORE), which employed 280 staff in May 1946, had a broad range of research, reporting, and estimating responsibilities, including economic intelligence.[2]

The evolution of the requisite CIA authorities for economic analysis over the next few years makes a fascinating case study in organizational behavior and public administration, but it is of only tangential concern here. What is important is that in June 1953, a five-person Military-Economics Branch was authorized for the Office of Research and Reports (ORR) in CIA. Its assigned function was to estimate the costs of the Soviet military establishment as an integral part of the larger ORR effort to estimate Soviet GNP.

It was not until a year later, in September 1954, that CIA was formally tasked with responsibility for the "production of all economic intelligence on the Soviet Bloc"[3]—a function it had already been performing for a number of years. The same year, responsibility for production of military-economic intelligence for all foreign countries was assigned not to CIA, but to the Department of Defense (DoD). The directive noted, however, that CIA "will supplement the intelligence produced by other agencies by conducting such independent analyses and studies as may be necessary to produce integrated economic intelligence on the Bloc." Clearly, an understanding of the size of the Soviet military-economic effort was central to the production of "integrated economic intelligence on the Bloc." The charter for the CIA role in the production of military-economic analysis on the Soviet Union was thereby indirectly but firmly established.[4]

## *How to Estimate Soviet Defense Spending*

In the late 1940s and early 1950s, the USSR released on an unpredictable schedule fragmentary information about its military budgets. After the merger of the Soviet War and Naval Ministries into the Ministry of Defense in 1955, the government established the routine of officially announcing a single figure for defense spending as part of the annual Soviet budget released by the Minister of Finance. This single figure, despite being part of an official economic document, was usually embedded in a context of political propaganda that raised questions about its reliability. (In contrast, for decades the DoD section of the annual U.S. budget document has consisted of one hundred or more pages of detailed tables, definitions, and program descriptions.)

Faced with this lack of official information on the Soviet military budget, CIA analysts needed methods for independently estimating Soviet defense spending. First, they needed to establish more confidence in the level of total Soviet military spending than the single figure provided. Equally important, national security policymakers and planners needed an understanding of the programmatic composition of the Soviet defense budget.

Despite the fact that they had a thin reed on which to build, the analysts of the late 1940s and early 1950s initially accepted as a working hypothesis the proposition that the announced budget covered almost all Soviet military activities. They believed the figure excluded spending for nuclear warheads, militarized security forces, and probably some military research, development, testing, and evaluation (RDT&E). In light of later developments that revealed the bogus nature of the announced budget, this early analytic judgment seems naive. In fairness, the size, behavior, and information on World War II Soviet defense budgets, the available Soviet financial literature, and even some intercepted unclassified Soviet military communications all supported the initial working hypothesis about the scope and reliability of the budget. Indeed, it may well be that in the immediate postwar period, the announced budget was what it purported to be.

Two basic approaches for independently estimating Soviet military spending (and thereby testing the working hypothesis) rapidly emerged. One relied on exploiting pertinent available Soviet economic statistics.[5] The second eschewed the use of Soviet statistics and instead employed a direct-costing technique of putting price tags on known and estimated Soviet military forces, programs, and activities.

Although the first basic approach rejected or at least seriously questioned the announced defense budget's validity, it assumed that certain Soviet economic statistics not identified as military-related in fact covered both civil and military activities, and that these statistics were accurate. In other words, military elements were hidden in categories with nonmilitary titles. The idea was to glean insights from these data on the magnitudes of the unidentified military components they contained by netting civilian components from them. For obvious reasons this is known as the residual method. It is in essence a macro approach to the problem; it begins with an economic aggregate and ends with a subaggregate labeled as military.

The second general technique—direct costing—is a micro approach to the problem. It begins with the smallest identifiable pieces of military activity, estimates their costs, and then combines the results into analytically useful aggregates. It is a straightforward approach: price *(p)* times quantity *(q)* equals spending; then spending for individual items is summed into subtotals and totals. Again, for obvious reasons it is known as the building-block method. The building-block approach became the method of choice for CIA military-economic analysis early on and served as the basis of official CIA estimates throughout the agency's work on Soviet defense spending.

CIA, however never entirely abandoned the residual method. For many years estimates of Soviet spending for military RDT&E, a major element of the estimates of total Soviet defense spending, were based on a residual methodology. And CIA analysts used residual techniques periodically as rough checks on the reasonableness of the building-block results. The Defense Intelligence Agency (DIA) in the DoD also relied on a residual technique for some of its analysis, and it was the principal method academic and other analysts outside the intelligence community used to estimate Soviet defense spending.[6]

CIA's choice of the building-block approach as its basic methodology in the mid-1950s was largely a matter of necessity. Those in charge of the effort believed that the scarcity of pertinent, reliable Soviet economic data created the need for a methodology that did not rely on such data—particularly in light of the Soviet Union's obvious attempt to conceal as much as possible about the military aspects of its economic activity.[7] Even after 1956, when the USSR began releasing substantially more information about its economic performance, the military sector remained shrouded in a fog of secrecy and definitional ambiguities.

Although the nature of the evidence may have been the driving force in the decision to pursue the building-block methodology, the approach was also highly compatible with the prevailing analytic culture of ORR. Dr. Max Millikan of the Massachusetts Institute of Technology, the first director of ORR, articulated his definition of economic intelligence as "intelligence relating to the basic productive resources of an area or political unit, the goals and objectives which those in control of the resources wish them to serve, and the ways in which and the effectiveness with which these resources are in fact allocated in the service of these various goals."

Millikan listed five objectives of economic intelligence, and the first four are concerned with military-economic analysis. They are as follows: to estimate the magnitude of possible present or future threats to the United States and its allies by evaluating the total economic resources on which the possible enemy's military potential must depend; to estimate the character and location of such threats by learning the potential enemy's allocation of these resources; to assist in estimating the intentions of the potential enemy; and to aid in decisions on policies and actions that can reduce the possible threat by seizing on or creating economic vulnerabilities.[8]

Millikan's definition and objectives of economic intelligence not only reflect a keen appreciation of the need for the policy relevance of economic intelligence, they also effectively demand a microanalytic approach to answering policy questions. They imply a highly detailed understanding of the inner workings of an economy that can be achieved only by starting at the bottom and working up. The analytic modus operandi Millikan suggested was to define the research objective; to construct what he called an "inventory of ignorance" by identifying all the things one needed to know to reach the objective, but did not know; and to reduce the unknowns through a continuing process of "successive approximations."

Although Millikan served only a little more than a year (January 1951 to March 1952) as the first director of ORR, his concepts about the nature and role of economic intelligence had a profound and lasting impact on the organization, staffing, and research planning of the office, particularly its military-economic effort. The conventional academic approach to estimating Soviet defense spending would have been to assign an analyst or two to the task of monitoring and exploiting Soviet statistics. Such a statistical monitoring effort, of course, did occur, but—given the sparseness and questionable reliability of Soviet economic statistics—it was recognized from the outset that more than official Soviet statistics would be needed to provide useful, reliable insights into the Soviet military effort. Only a Millikan-style approach of mastering the nitty-gritty details would do the job. In other words, a building-block method was needed.

## The Building-Block Method: Some Important Implications

Given the magnitude of the analytic effort required by the building-block method, the decision to use this approach is a measure of the importance of Soviet defense spending to U.S. national security policy and of the ORR leadership's confidence in its analytic capabilities.

In retrospect, we find the research and analysis task overwhelming. The first step required constructing a comprehensive articulation of the entire Soviet military establishment and its supporting armaments production programs into distinct cost-generating entities. The second step required the determination of an appropriate price to apply to each cost-generating entity. Indeed, it soon became clear that the second step would have to be done twice—once with a ruble price to estimate the expenditures as seen by the Soviets, and then again with a dollar price to show what a particular Soviet item would cost in the United States.[9] Finally, the results had to be summed and organized into an analytically useful structure. Despite this large effort, the building-block method offered important advantages.

WHO CALLS THE SHOTS? From the viewpoint of early ORR managers, one advantage of the building-block estimates was that they would be largely independent of Soviet statistics. The definitions, scope, and concepts associated with the spending estimates would be under the control of U.S. analysts, not Soviet leaders and propagandists who could—and did—change the meaning of their announced statistics without notice or explanation. Knowing exactly what one is measuring creates a solid foundation for responsible analysis and is a prerequisite for confidence in the results.

For the early building-block estimates, the guiding principle for defining the scope of Soviet defense spending (or Soviet military spending) was to make it the equivalent of the term "major national security expenditures" used in the official U.S. budget. Estimates for the USSR thereby covered outlays for equipping, operating, and maintaining the active military establishment, including the militarized security forces; pay and subsistence of reserve forces on active duty status; military pensions; RDT&E; and the production of nuclear weapons, which at the time in the United States were funded by the Atomic Energy Commission rather than the DoD. Although minor changes were made over the years in the scope of the estimates, the guiding principle remained: to match as closely as possible the U.S. concept of major national security expenditures. (In the late 1970s and early 1980s additional estimates were made, employing a substantially broader definition of defense spending in order to assess the full "cost of empire" to the USSR. We discuss these in chapter 5.)

The ability to tailor statistical definitions to match either Soviet or U.S. military and economic concepts made the estimates useful to a broad array of planners and policymakers. Costing the Soviet effort piece by piece permitted analysts to add expenditure pieces in several ways. Tables 2.1 through 2.3 provide samples in summary form of the taxonomies CIA developed to categorize Soviet military spending. Most of the categories shown were further disaggregated—in some cases several times—into subsidiary elements. In analyst working papers completed in the late 1950s, for example,

## TABLE 2.1
### SPENDING ESTIMATES BY ECONOMIC RESOURCE CATEGORY

Personnel
    Pay and allowances
    Food
    Clothing
    Medical
Operations and Maintenance
    Operating spares
    Alterations and maintenance
    Petroleum, oil, and lubricants (POL) consumption
    Maintenance of facilities
    Civilian personnel expenses
    Communications
    Transportation
Procurement of Equipment
    Land armaments
    Naval ships and boats
    Aircraft
    Missiles
    Electronic equipment
    Ammunition
    General purpose vehicles
    Organizational equipment
Construction of Facilities
    Airfields
    Naval bases
    Missile sites
    Personnel facilities
Research Development, Test and Evaluation (RDT&E)
Other

the military forces shown in table 2.2 were further broken down into approximately 260 types of military units for costing purposes. By the 1990s this number had increased to almost 1,800.

Expenditures for the resource categories that make up the total military pie—personnel, procurement of hardware, operation and maintenance

**TABLE 2.2**
**SPENDING ESTIMATES BY MILITARY SERVICE**

Ground forces
Naval forces
Air forces
Strategic rocket forces
Militarized security forces
Other forces

of equipment and facilities, new construction, and RDT&E—are of particular interest to those concerned with the economic impact of defense activities. Expenditures organized by each branch of military service—e.g., ground forces, naval forces—are of particular interest to analysts of Soviet bureaucratic behavior and internal politics. Expenditures organized by military mission—e.g., strategic offense and defense—are more likely to be of concern to national security analysts and planners.[10] Combinations and permutations of the taxonomies can also be constructed to address specific questions. For example, using a building-block method, it is possible to address questions like, "What are the estimated annual personnel expenditures for the air force elements of the strategic offensive mission?"

ROOM TO GROW. The building-block method offered another advantage over the residual method: almost limitless opportunities for refining the estimates over time (Millikan's successive approximations). At first, when the supply of pertinent intelligence information was sparse, only rough estimates could be made, usually at fairly high levels of aggregation. As the quality of the underlying evidentiary base of quantities and prices improved, estimates could be made at increasingly lower levels of aggregation, providing greater specificity and confidence. Initially, available intelligence information, for example, did not support an estimate of annual Soviet ground forces personnel expenditures more precise than that which could be made by multiplying an estimated average annual personnel cost per person by the estimated total size of the ground forces. As the information base improved, it became possible to calculate separately the annual personnel bill for officers, reenlistees, and conscripts. A further refinement permitted organizational breakdowns of the estimates; this made it possible to distinguish, for example, the annual personnel costs for a motorized rifle division

Soviet Defense Spending

TABLE 2.3
SPENDING ESTIMATES BY MILITARY MISSION

---

    Strategic Offensive Forces
        Intercontinental attack forces
            Long-range bombers
            Ballistic missile submarines
            Long-range missiles
        Peripheral attack forces
            Medium-range bombers
            Missile submarines
            Intermediate/medium-range missiles
    Strategic Defensive Forces
        Fighter air defense forces
        Surface-to-air missile forces
        Control and warning forces
    Conventional General Purpose Forces
        Ground forces
        Naval forces
        Tactical air forces
        Military transport forces
    Command and Support Forces
    Research, Development, Test and Evaluation (RDT&E)

---

from those of a tank division, or an artillery battalion from an infantry regiment. While this description makes the course of actual events seem much simpler than it was, it is in fact what occurred over the forty years of development of CIA's building-block method. The taxonomy of spending provided by the resource category breakdown constituted the foundation for continuing refinement and assessment of the uncertainties in the database.

Two factors account for the selection of the resource category taxonomy as the organizing structure for research. In the 1950s when the effort began, it was the primary way U.S. defense expenditures were organized, and for comparison purposes it made sense to set up Soviet spending accounts with the same definitions. Even more important, it fit nicely with the bits of intelligence information on Soviet military spending that were gathered. Individual pay rates for various military ranks and jobs,

**Getting Started**

costs of specific military rations, prices of individual pieces of military equipment, and construction costs for various military facilities typify the information available to analysts.

The specific and detailed answers that the building-block method can provide are important payoffs. The other side of the coin is that zeros simply are not allowed in the mathematical process employed by the building-block method. All information gaps must be filled. There is little question that in the early days the "inventory of ignorance" was substantially larger than the inventory of known facts. We described one way to deal with these gaps in the discussion of estimating personnel costs for the Soviet ground forces—that is, the aggregation technique of moving up the echelons of detail for which information is not available until an aggregation level is reached that can be filled with data of an acceptable confidence level. In the example cited, a single average annual personnel cost figure was used until more detailed information could be obtained. Another technique, which was more desirable because it did not sacrifice detail, involved interpolation or extrapolation of known information. For example, if the monthly rank pay of a Soviet army major was known to be $x$ rubles and the pay for a colonel was known to be $y$ rubles, but the pay for a lieutenant colonel was unknown, interpolation permitted an analyst to fill the gap fairly reliably with a figure somewhere between $x$ and $y$.

RELIANCE ON U.S. ANALOGS. A less desirable method for filling information gaps was the use of U.S. analogy; by definition such information was a step removed from Soviet reality. On the other hand, U.S. analogy was better than a wild guess and, if intelligently applied—"Sovietized"—it provided a guide to establishing a reasonable cost estimate for a Soviet military item or activity similar to a U.S. military item or activity. It is important to note that—contrary to the charges of some critics of CIA estimates—analogous U.S. information was not used to develop broad macro judgments about Soviet military spending levels or patterns.[11]

It is clear from reviewing analyst working files and methodology papers of the early period that while the analysts used U.S. analog information extensively, they were aware of the pitfalls of mindless "mirror imaging." To minimize the likelihood of introducing false analogies into the building-block framework, analogs were usually applied at the lowest possible level of detail; here they were more likely to provide appropriate matches and their reasonableness could be assessed more easily. In other words, the same microanalytic approach used to exploit direct Soviet data was applied to U.S. analog data.

In the 1950s analogs were used extensively to fill information gaps about the more mundane, but nonetheless important, cost-generating items. These included purchases of miscellaneous supplies and equipment—general purpose vehicles, field kitchens, tents, and office supplies, as well as the nonpersonnel daily expenses of operating and maintaining the military force structure (fuel and lubricants, spare parts, equipment overhaul and maintenance, transportation, and communications). Often the analog served as a point of departure in developing an estimate and was not applied directly in unmodified form. For example, if the annual fuel costs of a U.S. diesel submarine were used to estimate the annual dollar fuel costs for a counterpart Soviet submarine, the U.S. cost data would first be modified to reflect known differences in the propulsion systems and the average number of days at sea for the two classes of submarines.

Analogs also played an important role in the all-important area of estimating outlays for procurement of major weapons systems. Soviet price information for items such as tanks, artillery, aircraft, ships, and missiles was scarce.[12] And although Soviet civilian analog information—for example, merchant shipbuilding costs—provided significant insights about costs in Soviet military industry, U.S. analogy provided the essential foundation for estimating a substantial share of Soviet expenditures for procurement of military hardware. Again, more often than not a pure U.S. analog was not used to estimate the cost of a counterpart Soviet system; the U.S. cost was first modified to reflect known and estimated differences in the performance characteristics and engineering specifications of the two systems. As the body of information on the characteristics of Soviet systems increased, the need to use cost data for counterpart U.S. systems decreased. If enough was known about a Soviet weapon, the cost of the analogous U.S. weapon could be skipped in the estimating process and a dollar cost for the Soviet weapon could be estimated directly.

It was ideal to have possession of the item itself or detailed manuals and plans of the item.[13] In either case a dollar cost for manufacturing the Soviet item could be determined directly by a U.S. manufacturer. U.S. manufacturers also developed cost-estimating relationships (CERs) that made it possible to estimate the production costs of specific items of military hardware, based solely on key performance and engineering parameters. Obviously the more that was known about these parameters, the more detailed and accurate the CER was likely to be. The great advantage of direct- and CER-costing techniques was that the price tag for a Soviet item was based directly on that item; it was not the price of a U.S. system that resembled the Soviet item or that wasn't much like it at all but served the same function.

We cannot easily determine the extent to which early CIA estimates of Soviet procurement expenditures relied on U.S. analogs. A detailed study published in 1988, however, shows that by that time about 30 percent of the total was analog based and about 70 percent was based on more direct methods.[14] Given the substantial amount of system cost analysis done in the intervening thirty years—particularly in the late 1970s and the 1980s—a reasonable guess would be that in the 1950s 50 percent or more of the procurement estimates was based on U.S. analog information.

The use of U.S. analogs in the costing analysis helped focus military-economic analysts' and intelligence collectors' attention on important information gaps bearing on the readiness of the Soviet forces and the capabilities of their weapons. Knowing precisely how the Soviets trained and how they utilized their equipment—e.g., flying hours for aircraft, days at sea for ships—had a significant impact on estimates of operating costs. Similarly, the procurement costs of weapons are frequently highly sensitive to small variations in key characteristics and performance parameters. A synergistic relationship developed between the military-economic analysts and the military analysts, who were charged with estimating the capabilities and readiness of Soviet forces. The push to replace analog costing with direct costing frequently added enough priority to the collection requirements to ensure an effective effort to fill gaps in CIA's knowledge of Soviet weapons characteristics.

## Monetary Measures—Rubles and Dollars[15]

The fact that some of the basic cost data collected to feed the building-block methodology was expressed in rubles (for example, military pay rates) and some was expressed in dollars (for example, U.S. analog costs) created the necessity for a way to express all elements of spending in a consistent currency. Overshadowing this purely technical requirement for a valid conversion mechanism was the conceptual imperative for both a ruble and a dollar valuation of Soviet defense activity. As we noted in chapter 1, the two enduring fundamental questions policymakers posed about the Soviet defense effort were its size and the burden it imposed on the Soviet economy. Unfortunately, these questions cannot be addressed with any single monetary measure.

## The Sizing Problem

The USSR spent rubles, not dollars, on its defense programs. But CIA estimates of Soviet defense spending, expressed in rubles, conveyed little if any appreciation of the size of the Soviet defense effort to planners or policy officers in the national security community, or to Congress. Telling the Chief of Naval Operations or the chair of the House Armed Services Committee that the USSR was spending 15 billion rubles on its navy was likely to evoke a response like, "What the hell does that mean? It doesn't tell me anything about how big, how good, or how active their navy is!" On the other hand, telling them that the estimated dollar equivalent figure was about $20 billion struck a chord; they knew from their own experiences what kinds of naval programs $20 billion could buy.[16] And because the dollar figures were the product of the building-block methodology, they were particularly meaningful to these consumers because they could be presented at detailed levels of aggregation and in the same accounting structure as U.S. defense spending. As such, they provided a convenient and legitimate basis for direct comparisons between the size of a given Soviet defense activity and a similar U.S. activity. The requirement for dollar valuations of the Soviet defense effort was born.

We should be clear about the conceptual meaning of the CIA dollar valuations of Soviet defense activity. The 1988 CIA study put it this way:

> *The dollar value of the defense activities of a foreign country is calculated by estimating the cost of each activity in the prices and wages prevailing in the United States . . . each activity is directly costed in U.S. prices and wages. For example, the dollar cost assigned to a year's production of T-80 tanks in the Soviet Union would be found by estimating the cost to produce the same tanks, at Soviet production rates, using U.S. material and labor costs and U.S. manufacturing practices—and, therefore, U.S. manufacturing efficiencies. The dollar cost of maintaining the Soviet Union's large inventory of tanks for a year would be found by applying U.S. prices and wages to the Soviets' maintenance practices . . . personnel costs are directly calculated at U.S. rates.*[17]

In short, CIA's dollar figures attempt to show what it would cost to buy the actual Soviet military program in the United States. Being able to express Soviet and U.S. military activities in terms of a common denominator (dollars) provided a powerful analytic tool for national security policy officers. The use of dollars to size Soviet defense programs, however, was also an important source of criticism of CIA military-economic efforts. Some of these criticisms raise valid technical issues such as the index number problem, which is inherent in all international economic comparisons expressed in monetary terms. We discuss these technical issues in chapter 6.

Other criticisms stem from a simple failure to adequately grasp the conceptual core of the dollar valuations—that is, that they are no more and no less than estimates of what it would cost in the United States to replicate estimated Soviet programs. This is a straightforward concept, and we find it puzzling that it was so often misunderstood, especially in light of CIA's effort from the beginning to make it clear.[18] Nonetheless, misunderstanding existed and still exists. One former senior official, whom we interviewed for this book, told us that he distrusted CIA's dollar defense numbers primarily because the nonmarket nature of the Soviet economy distorted true economic relationships and made the dollar valuations meaningless. He apparently did not understand that the nature of the Soviet economy had absolutely no methodological connection with the act of applying dollar prices to physical descriptions of Soviet military forces and programs.

The most common criticism takes this generic form: "One has only to note that when the U.S. military gets a pay raise, CIA's dollar estimates of Soviet defense spending go up, and it becomes clear that there is something seriously wrong with the CIA numbers." As we have pointed out, the Soviets spent rubles, not dollars, for their defense program, so CIA's estimates of Soviet defense spending do not increase when the U.S. military gets a pay raise. The dollar valuations of Soviet defense programs, however, would go up as a consequence of the U.S. military pay raise, and rightly so. It would take more dollars to replicate the Soviet programs in the United States after the pay raise than before the raise.

## The Burden of Defense

The economic burden of Soviet defense programs is not a simple, clear-cut concept. It can be viewed from various perspectives with different forms of measurement. One objective of CIA's burden analysis was to transcend the distortions of the Soviet command economy and measure the real diversion of factors of production from the rest of the economy to the military sector. Another distinct goal was to understand how Soviet political, military, and economic leaders and planners might view the economic implications of their military spending and the trade-offs of alternative programs, given the realities of their economic system.

CIA military-economic analyses reflecting these perspectives required ruble estimates of military spending as the only valid basis for judgment. Dollar valuations cannot be used because they reflect the realities of U.S. economic scarcity relationships, and it is the Soviet economic scarcity relationships that determined the economic burden. These differences in internal economic relationships in the two countries were manifested in the countries' markedly different internal price structures. The structures' potential impact becomes apparent when one considers that, generally, Soviet outlays for procurement of equipment averaged roughly one-half of total Soviet defense spending when measured in rubles, but only about one-quarter when expressed in dollars. The bottom line is that from the very beginning of CIA's military-economic work, no single monetary measure could be used to legitimately address the full range of economic questions about Soviet defense activities.[19]

## *Dollar-to-Ruble Conversion Ratios*

Two full sets of prices would be needed to value the quantities describing Soviet military forces and programs. Again, the building-block framework made it possible to take advantage of the most reliable price information (ruble or dollar) at any point in time. Reliable price data not only filled the information gaps for specific items to which they applied, but also in many cases provided the basis for constructing dollar-to-ruble conversion ratios that made it possible to fill gaps in price information for a whole class of comparable items.

For example, if a reliable ruble price for a specific model of Soviet tank was obtained and enough was known about its physical and performance characteristics to develop a dollar cost for producing the tank in the United States, then an implicit dollar-to-ruble conversion ratio was created. This ratio could be used to derive ruble prices for other Soviet tanks—for which no ruble information was available—by applying the implicit ratio to the estimated dollar production costs in the United States.

We cannot emphasize enough the importance of valid dollar-to-ruble conversion factors in the defense-spending estimating process. A dollar-to-ruble conversion ratio, based on a single piece of price information, could exert substantial leverage over the spending estimate for an entire type or class of equipment or service. CIA recognized the ratio's importance in the early days of building-block methodology. Throughout the 1950s the agency placed a premium on collecting ruble prices and exploiting them through extrapolation, interpolation, and successive approximations—with an eye toward developing a database of dollar-to-ruble ratios that could help fill information gaps regarding actual Soviet ruble military prices.

The task's importance was exceeded only by its size and difficulty. It could not be accomplished effectively by the small number of military-economic analysts working in ORR. The job was simply too big. Fortunately, however, ORR's work on Soviet industry and economic infrastructure also required reliable dollar-to-ruble conversion ratios for comparisons of GNP and its components. Much of this work could be directly or indirectly used to support the military-economic analytic effort. Moreover, think tanks such as the RAND Corporation and academic researchers made important contributions to the development of conversion ratios through their work on Soviet civilian analog prices.[20] Consequently, ORR military-economic analysts were generally satisfied by the mid- to late 1950s that, with the exception of RDT&E, they had a useable database of Soviet military prices at their disposal, even though many of the military equipment procurement prices were indirectly estimated through the use of dollar-to-ruble conversion ratios.

The overall confidence of the military-economic analysts at the time was bolstered by the fact that the indirect route to Soviet military prices via conversion factors was not necessary for a substantial portion of the total package of goods and services purchased by the Ministry of Defense. All outlays for military personnel, for example, were estimated directly in rubles based on abundant information available on Soviet military pay rates—despite the fact that they were classified as state secrets—and food and clothing prices. Similarly, analysts had access to direct ruble price information for construction activities and major operating costs like purchases of fuel and lubricants. By the early 1960s approximately one-half of total spending was directly estimated in rubles without the need to rely on dollar-to-ruble conversion ratios.

Even if the dollar-to-ruble price relationships established in the mid-1950s were an accurate reflection of reality at that time,[21] it was unlikely that those relationships would remain the same over any extended period of time. True, the centrally controlled prices in the USSR were only officially adjusted every several years to reflect changes in the underlying economic phenomena, but on the U.S. side prices changed continuously to reflect shifts in technology, production efficiencies, and costs and availabilities of material and labor inputs. A ratio that was valid for 1955 dollars and rubles was unlikely to be valid for 1956 dollars and 1955 rubles, and even less so for 1960 dollars and 1955 rubles.

Another major analytic task was thus defined: close monitoring of the relationship between ruble and dollar price behavior over time would be required—in addition to collecting as many direct ruble military prices as possible—to refine and maintain the integrity of the database of dollar-to-ruble conversion ratios. Unfortunately, analysts who might be expected to focus on continuing price analyses were often diverted to other priority tasks. Managers, having made a substantial investment in building a price database, took the calculated risk of living off existing capital for a few years while they turned research attention to other problems. As it turned out, the general neglect of the ruble price database lasted far longer than planned, and its full impact on CIA ruble spending estimates did not begin to emerge until the early 1970s, when substantial new collection and research efforts on Soviet military prices were initiated.

Establishing accurate price tags for the hundreds of cost-generating items and activities in the total Soviet military effort required three distinct research tasks: obtaining direct ruble prices for military goods and services; developing accurate dollar costs for replicating Soviet defense activities in the United States; and maintaining the reliability of dollar-to-ruble price relationships over time.

GETTING THE QUANTITIES RIGHT. The building-block formula for estimating Soviet defense spending was simple, at least in concept: *prices x quantities = spending*. Getting the prices right was a tough job, but by the mid-1950s the managers of CIA's military-economic effort had enough confidence in their research to move away from use of officially announced Soviet data and rely on the building-block method for official CIA estimates.

Obtaining reliable data for the quantities portion of the formula should have been relatively easy for CIA analysts, but in fact it turned out to be even more difficult, at least bureaucratically, than the price problem. After all, determining the physical dimensions of the Soviet military threat to the United States and its allies was the primary focus of a large share of the intelligence community during the Cold War. The quantitative estimates on numbers of units in the order-of-battle, manpower levels, and production rates for weapons and equipment were precisely the data needed to complete the quantities portion of the formula. It should have been a simple proposition for the CIA military-economic analysts to plug the available estimates of quantities made by other elements of the intelligence community into the formula and calculate estimated spending levels, but it wasn't.

## The Charter Problem and the Credibility Problem

Before the creation of ORR and the Office of National Estimates (ONE) in 1950, not a single organizational entity in the predecessor organization (ORE) had a name that suggested any involvement in research and analysis on Soviet military forces and programs. At the time, CIA relied on the U.S. military services for military intelligence on the USSR. A few years later, however, the structure of ORR included—in addition to the Military-Economics Branch—analytic branches with responsibilities for Soviet aircraft, missiles, shipbuilding, and weapons and ammunition. What had happened to explain this change?

The period leading up to and following the creation of ORR and ONE was one of major bureaucratic turmoil in the intelligence community.[22] Organizational identities needed to be established in fact as well as form. Roles, missions, and functions had to be sorted out. As part of this general bureaucratic process, ORR faced a difficult dilemma in its formative years. To carry out its mission to produce reliable intelligence on the Soviet economy, it needed to understand the military demands on the economy, which in turn required accurate estimates of the underlying military forces and programs. But the military service intelligence organizations—which had a clear charter for military intelligence analysis—were not providing physical data that met ORR's requirements in terms of credibility or scope. ORR could either fail to do the best possible job or it could—under existing directives that permitted any analytic activity necessary to satisfy internal needs—venture into the realm of military intelligence. It chose the latter course and rapidly developed an independent capability to augment and, where necessary, challenge the military services' intelligence product.

On the issue of credibility, military service intelligence organizations' penchant for applying a worst-case scenario to estimate Soviet military capabilities and their acute sensitivity to the programmatic and budgetary goals of their parent services were probably the primary motivating factors in the creation of a centralized intelligence function under civilian control.[23] When combined with the knowledge of hindsight, a review of military service intelligence chiefs' dissenting footnotes appearing in NIEs on Soviet military programs provides ample, vivid documentation of the credibility problem. Recently, NIEs through 1984 have been declassified and deposited in the U.S. National Archives. The problem has also been widely discussed in intelligence literature.[24]

On the issue of scope, CIA military-economic analysts' and military intelligence analysts' priorities often diverged on the needs for quantitative information. In addition to high-profile and threat-generating weapon systems such as tanks, aircraft, submarines, and missiles, military-economic analysts were interested in unglamorous but substantial cost-generating items such as purchases and operation of trucks and other general purpose vehicles. These items often fell below military intelligence analysts' interest threshold, as their work was driven by very different imperatives.

The motives for CIA's early venture into military intelligence analysis extended beyond the drive to meet ORR's responsibilities for production of economic intelligence on the USSR. Early on, ONE, which had the responsibility for implementing the agency's assigned duty under Section 102 of the National Security Act of 1947 to "correlate and *evaluate* intelligence relating to the national security" (emphasis added), recognized flaws in the contributions of military service intelligence organizations to the NIE process and looked to CIA's analytic assets in ORR, the Office of Scientific Intelligence (OSI), and other analytic components of the Directorate of Intelligence (DI) for support in assessing the Soviet military threat. An internal, unsigned ORR memorandum dated June 6, 1952 provides an excellent example. The memorandum, prepared after consultation between ONE and ORR, assigns responsibilities within ORR for contributions to the document NIE-65, *Soviet War Potential, 1952–1957*. It notes, "we have a major responsibility to support and *check* the military agencies' contributions . . . [on] sizes and quality of holdings of military equipment by Soviet Bloc military forces, estimated size, quality and disposition of military stockpiles, military consumption rates of Bloc forces under cold and hot war conditions" (emphasis added).

Although the full story of the development of CIA's military analytic capability and the mix of motivations for it are not central to this study, the impact of the diversion of analytic resources from military-economic analysis to military analysis is.[25] In 1955 ORR had almost fifty positions putatively assigned to work on military-economic analysis, but only about ten of these positions involved research on military prices and cost factors, constructing estimates of defense spending, and assessing the economic impact of defense spending. The remaining positions were devoted primarily to work on the quantitative dimensions of Soviet military programs—for example, order-of-battle, military production, and military manpower levels. Some of this quantitative work filled gaps in the work of military intelligence services, but most of it focused on the same intelligence objective—that is, to determine the physical dimensions of the Soviet military threat in terms of quantity and quality.

The price ORR paid for entry into military analysis was high in terms of analyst resources diverted from direct military-economic work, but the benefits to the general intelligence process more than made up for the costs. The deep involvement of ORR and other CIA analytic components ensured that the system was working as the legislative designers intended and that the resulting assessments of Soviet military programs reflected a thorough review and evaluation of all pertinent information and analyses available in the intelligence community.[26] The payoff to intelligence consumers was obvious: they were getting the best judgments of the intelligence community as a whole. There was also an indirect payoff to ORR: it was getting a physical foundation for its methodology that reflected the combined efforts of all elements of the intelligence community.

An additional diversion of research attention from work on military prices and cost factors became increasingly onerous during the last half of the 1950s. As the building-block methodology took hold and more detailed information was collected, the purely mechanical and computational tasks of constructing spending estimates grew astronomically. By the late 1950s tens of thousands of individual calculations were involved, and they all had to be structured into coherent form. Today, with computers readily available to every analyst, the task would be simple, but in the 1950s, when slow electromechanical desk calculators were the order of the day, the task created a significant challenge in the effective management of resources. Chapter 3 discusses this problem further, but we note here that the conceptual challenges and computational tasks involved in constructing the expenditure estimates continued to require substantial analytic effort.

## The Early Estimates

### Reflections in the NIE

The first few NIEs dealing with Soviet capabilities and intentions published after the formation of ORR and ONE in 1950 contained no explicit numerical estimates of Soviet defense spending. The only pertinent reference in NIE-3, issued five months after North Korea's invasion of South Korea in June 1950, is that "The Soviet Union has already largely mobilized its industry for war."[27] By November 1952, NIE-64 contained a more expansive discussion on the size and growth of the Soviet economy and included the following estimative judgments about defense spending: "We believe that the USSR now devotes about one-fifth of its national product to military expenditures; and [we] estimate that the quantity of resources which the USSR will devote to military production in 1952 will equal the amount assigned in 1944."[28] Again, no specific numerical estimates for either GNP or defense spending were presented. Just over six months later, in June 1953, NIE-65 stated, "We estimate that about one-sixth of Soviet gross national product is now devoted to military outlays."[29] NIE 11-4-54, issued in September 1954, states: "Current economic programs indicate that for at least the next two years the amount of expenditures on defense, instead of continuing the rapid increase that prevailed in 1950–1952 will remain about the same . . . ; and [the] pattern of resource allocation in the Soviet economy in 1953 showed about 14 percent devoted to defense."[30]

In addition to the lack of specific numerical estimates for defense spending, other common elements in these early statements are worth noting. First, the nature and context of the judgments indicate that they were derived from a macroanalysis of the Soviet economy, based largely on officially announced Soviet statistics. Second, they do not indicate any significant economic constraints on Soviet military programs; the shares of GNP cited are substantial, but they decline from a 1952 estimate of 20 percent to a 1954 estimate of 14 percent while the overall economy was estimated to be growing rapidly.

The issuance of NIE 11-6-54, *Soviet Capabilities and Probable Programs in the Guided Missile Field,* in October 1954 completely changed the situation. One of the major conclusions of the NIE was: "The USSR . . . has an adequate economic base for a sizable missile production program. However, because of the limited capabilities of the Soviet electronics and precision mechanisms industries and other competing demands for their output, the USSR will almost certainly be unable to produce in the desired quantities all of the missiles for which it has an estimated military requirement, except over an extended period of years."[31] This conclusion is supported by a two-page discussion of the evidence and methodology later in the body of the NIE, as well as by an eight-page annex of detailed tables—including estimated costs—and text.

Getting Started

The building-block method for estimating military expenditures had been used for the first time in the national estimating process, and it had a major impact on a key NIE's conclusions. There was still a long way to go before the building-block method could be extended to the total Soviet military establishment, but the concentrated effort on the critically important, but relatively small, guided missile program vividly demonstrated the utility of the approach to military spending analysis. The same kind of analysis was conducted in the spring of 1955 in support of NIE 11-5-55, *Air Defenses of the Sino-Soviet Bloc, 1955–1960*. A CIA/ORR contribution to the NIE[32] concluded on the basis of economic analyses that the production and deployment numbers postulated for inclusion in the NIE, particularly those for fighter aircraft and guided missiles, were unrealistically high:

1. *Considering other military requirements, implementation of the air defense program postulated . . . would require a considerable rise in military expenditures.*
2. *This program . . . would utilize so much industrial capacity that it would reduce the rate of growth of heavy industry.*
3. *[I]mportant portions of the program . . . such as aircraft and guided missiles, could be fulfilled on the required scale . . . only if the capability of the Soviet Bloc electronic equipment industry is expanded to a far greater extent than is currently estimated.*
4. *Since the air defense program postulated . . . is so costly in terms of industrial output in general and electronic equipment in particular and since it would require a sizable increase in defense expenditures, the Soviets will have to be satisfied with fewer production models of aircraft and missiles and some reduction in the provision for AAA fire control.*[33]

This analysis had a substantial impact on the final version of the NIE, which included a four-page section in the main body under the heading, "Economic Impact of Air Defense Program," supported by three additional pages of detail in an appendix.[34]

### Solidifying the Building-Block Approach

According to John G. "Jerry" Godaire, the first chief of the Military-Economics Branch, both Sherman Kent, director of ONE, and Robert Komer, chief of the ONE estimates staff, were pleased with ORR efforts and recognized that the building-block costing analysis had the potential to be a unique and important tool for testing the reasonableness of estimated Soviet military programs, both individually and in the aggregate.[35] They also recognized that the costing discipline was in its infancy, and they did their best to promote its development. Soon thereafter, perhaps as a result of ONE's interest, the Military-Economics Branch was increased from five to eleven positions. But the most significant event in the development of the building-block costing methodology was ONE's issuance in August 1955 of a "terms of reference" memorandum for the preparation of NIE 11-4-56, *Soviet Capabilities and Probable Courses of Action through 1961*. The terms of reference called for a comprehensive and detailed building-block cost study, and an interagency committee under CIA chairmanship was created to ensure completion of the study.[36]

Although we were unable to document it from the written record, there were probably several reasons for commissioning an interagency group to prepare the first comprehensive building-block estimate instead of tasking CIA/ORR with the job. The quantitative information on forces and programs required for the costing extended well beyond the more summary information usually provided by the military service intelligence units for NIEs. A working-level interagency mechanism was required to obtain the required data and demonstrate the active involvement of military services. There may have also been hope that the services' participation in the process might commit them to the results.[37] If this indeed was an objective, it was only partially achieved. The director of Naval Intelligence and the assistant chief of staff for Intelligence, Department of the Army, essentially disavowed the study's results in dissenting footnotes. Nonetheless, the military expenditure annex to NIE 11-4-56 firmly established the building-block approach as the method of choice in estimating total Soviet military expenditures and confirmed the practice of including consideration of the spending implications of estimated forces and programs in the preparation of pertinent NIEs for the remainder of the decade and beyond.

The results of the study were reported in a six-page annex to the NIE[38] and in summary judgment form in the text.[39] The committee reported three major purposes of the study: "The first is to measure over time the economic burden placed on the USSR by its defense establishment through calculation of a ruble total which can be compared with gross national product. The second is to examine these expenditures in detail in order to permit closer analysis of the composition and trends within this total. The third is to measure the dollar value of Soviet defense expenditures in order to obtain some comparison, albeit a crude one, of the value of defense expenditures in the USSR and the US."[40]

The first purpose, of course, was not new, but for the first time NIE judgments were based on a time series (1950–61) of total defense spending that used the building-block approach to directly reflect the intelligence community's assessment of the Soviet military establishment. Also, for the first time the NIE was able to present breakdowns of total spending into detailed spending categories. Finally, the micro approach was a major stride forward in the analytic effort to effectively convey to U.S. audiences the Soviet programs' size. An unpublished draft prepared by ORR members of the interagency committee to document the work done in support of the NIE had this to say: "the following useful goals have been attained: 1) coverage of estimates of Soviet military expenditures have been extended in depth and detail; 2) a method and a framework capable of assimilating virtually all pertinent considerations and data have been established; 3) a definite advance has been made in the economic measurement of estimated Soviet military programs and in the interpretation of the defense expenditures announced in the Soviet State Budget; and 4) a foundation for the improvement of future estimates has been laid."[41]

The NIE itself summarized the achievements of the interagency committee's analytic effort more modestly: "Because the analysis was conducted in far greater detail than previous work in this field and took into account many factors not previously examined, we believe that the results are considerably better grounded than our earlier estimates."[42] A formal postmortem analysis of NIE 11-4-56 issued by ONE after completion of the NIE rightly focused on ways to further improve the estimates: "The findings of the IAC Ad Hoc Military Cost Study Committee incorporated in NIE 11-4-56 represent a substantial advance in this field but still leave much to be desired in terms of definitiveness and an appreciation of margins of error involved . . . effort in this field can fruitfully be concentrated upon revision of physical estimates of manpower and procurement . . . upon increased price collection and the improvement of indirect pricing techniques; and upon basic work on such segments of the Soviet defense effort as research and development and atomic energy, the cost of which must now be estimated in the aggregate."[43]

For the remainder of the 1950s and into the early 1960s, ORR focused its military-economic program on building its capital stock of research on the military-expenditures estimating process along the lines suggested in the postmortem memorandum. In addition the office continued to provide direct support in the form of contributions on military spending to pertinent NIEs. Much of the research effort was documented in ORR research reports on subjects such as the Soviet military pay system, Soviet military manpower, ruble-to-dollar ratios, Soviet aircraft production costs, and Soviet shipbuilding costs.[44] The publication record reflects a substantial effort to refine the expenditure estimates. Three of the papers published near the end of the period merit particular attention because together they summarize the progress:

- Methodology for Estimating Soviet Military Expenditures, *August 20, 1960. The first fourteen pages of this ninety-seven-page research report provide a general conceptual-methodological statement explaining the building-block technique and its system of spending accounts. The remainder presents individual statements describing the derivation as of 1960 of both ruble and dollar estimates for each of the system's cost accounts.*
- Soviet Military Expenditures, 1958–1965, *April 1961. This eighty-two-page report uses the Soviet military forces and programs contained in NIE 11-4-60,* Main Trends in Soviet Capabilities and Policies, 1960–65, *as the quantitative base for a detailed presentation of the summary expenditure information that appears in the NIE. It is the first detailed intelligence analysis of Soviet defense spending based on the building-block technique to be broadly disseminated in the U.S. national security community. The expenditure information is presented in rubles and dollars in a format that matches the resource category structure of the U.S. budget. The paper also includes a discussion of the relationship between CIA estimates and the Soviet announced budgetary allocations for defense. It concludes, ". . . the announced allocations cannot be used safely as a benchmark for testing the reliability of the estimates."*[45]
- Soviet Military Expenditures, by Major Missions, *April 1961. This is the companion paper to the April 1961 paper cited above; the important distinction between the two is that this paper structures the expenditure information on a military mission basis, rather than a resource category basis, to provide a different perspective on Soviet defense programs. As such, it foreshadowed the primary focus of military analysis in the U.S. national security community in the early 1960s.*

# Chapter 3
# New Questions, New Techniques, and New Evidence, 1961–74

Sometime around midday in Washington, D.C., on January 19, 1961, it began snowing hard. By four in the afternoon the snow had accumulated so rapidly the federal government decided—too late—to simultaneously release all employees from work early. The result was traffic chaos. We happened to be in the same car pool at the time, and we arrived home at about two-thirty the next morning—the day of President Kennedy's inauguration. The arrival of the Kennedy administration was accompanied by another kind of blizzard and ensuing chaos—the rapid buildup of demands for military-economic analyses and the turmoil they created. Major improvements in satellite intelligence collection capabilities also contributed to rapid changes in the overall military-economic analytic environment. This chapter discusses these changes and the organizational responses to them.

## New Demands for Military Expenditure Analysis

### RAND and New Approaches to Defense Analysis

In his 1983 book, *The Wizards of Armageddon*, Fred Kaplan presents a fascinating chronicle of the post–World War II evolution of U.S. strategic thinking and defense analysis.[1] The RAND Corporation is featured prominently in Kaplan's book because it played such an important role in that evolution—both as the original defense think tank and as the source for defense intellectuals who moved into the Pentagon in the early 1960s. RAND also had a great impact on the evolution of military-economic analysis in CIA.

Long before Robert McNamara and the RAND "whiz kids" revolutionized the way the Pentagon did business, CIA's military-economic analysts and members of RAND's economic division under Charles Hitch had cultivated an active working relationship. In the 1950s, RAND was the only research entity besides CIA to devote a substantial effort to investigating Soviet defense costs—some of it on contract to the agency—and it shared an affinity for the microanalytic approach to problems implicit in CIA's building-block methodology. It seems likely, for example, that the decision by managers of CIA's military-economic effort in the late 1950s to organize the estimates of Soviet defense spending by military mission, in addition to the traditional structures of resource categories and military services, reflected a familiarity with and sympathy for RAND's functional approach to defense analysis.

This functional approach was the essence of the McNamara/whiz kid revolution: the annual defense budget was the battlefield, systems analysis and cost effectiveness analysis were the primary weapons, and numbers were the ammunition. The traditional institutional focus on the air force, navy, and army budgets did not totally disappear, but now it competed with a functional focus as the driving force. Programs had to be explained and justified in terms of military missions that cut across service lines—for example, Minuteman intercontinental ballistic missiles, (Air Force); B-52 bombers, (Air Force); and Polaris submarine-launched ballistic missiles, (Navy), all fell into the same functional mission of strategic offense.

## *McNamara's Pentagon and Its Impact on Military-Economic Analysis*

Quantification of the potential military threat to U.S. national security had always been important, but given the nature of early 1960s defense analysis in the Pentagon, the demand for "threat" numbers suddenly increased by several times. Even relatively detailed numerical data took on a new dimension of significance. A 10 percent difference in the estimated value of a key performance parameter of a Soviet system, for instance, might drive the outcome of the analysis performed in the Pentagon. Because the costs of U.S. programs played such a central role in the new Pentagon analysis, Soviet defense costs inevitably assumed new importance as part of the process.[2]

Some of the new demand for military-economic analysis came directly from former RAND employees familiar with CIA's work, including Charles Hitch, the new Pentagon comptroller, and Alain Enthoven, director of the new DoD systems analysis shop. Most of the new demand, however, developed indirectly as a result of the changes in the NIE process instituted to meet new defense strategists' and planners' needs. In early 1961 CIA Deputy Director for Intelligence (DDI) Robert Amory commissioned a RAND study to advise him on how to adjust the NIE process to maximize its responsiveness to the new defense policy process.

The result, *Project Lamp,* was submitted to the DDI in April 1961.[3] The authors list the study's objective: "As first priority, to examine the potential application of systems analysis techniques to the process of producing national intelligence estimates on the Soviet military posture."[4] The study called for projections of alternative Soviet force structures for the "midterm period," meaning five to ten years in the future. This proposal did not win universal support. Sherman Kent, director of ONE, commented: "we do not subscribe to the view that alternative projections would be the preferred method of handling near-term estimates. While this would provide military gamers with more variations to explore, we think the tendency would inevitably be to inject a 'worst case' philosophy into all estimating. We have a responsibility, not only to the military planner but especially to the political planner, to project the probable situation as far forward as the evidence justifies. The trick is to judge when this can be done for, say, five years or one year or not at all."[5]

Kent's view notwithstanding, the projection of alternative Soviet force structures did become part of the NIE process for the next several years, first in the form of something called "Intelligence Assumptions for Planning," first issued in 1964. Presumably the name was intended to distinguish this work from the NIEs on Soviet forces that were still produced. Over time some of this distinction was lost, and the document was renamed "National Intelligence Projections for Planning."[6] In addition to these national intelligence vehicles, other analytic avenues were explored. One such response was the Joint Analysis Group, a small number of senior CIA and DIA analysts whose function was to generate alternative Soviet force structures without subjection to constraints imposed by the coordinating mechanisms of the NIE process. The number of force combinations and permutations requiring the production of expenditure implications by CIA's military-economic analysts seemed endless, and it essentially took all available resources to keep up with current demand. Little time was left for maintaining the database of price and cost information underlying the costing process.

## DIA and Its Many Customers

Once DIA was established in 1961 to serve the intelligence needs of the secretary of defense and the joint staff, it is not surprising that many DoD components turned to DIA for military-economic intelligence support. Indeed, DIA tasking sometimes duplicated requests that had also been levied on CIA. Although it had competent economic analysts, DIA had neither the experience nor the databases CIA military-economic analysts had been building for nearly ten years. And it did not have anything like the building-block costing system at its disposal. DIA analysts were, of course, aware of CIA's detailed costing work in support of the NIEs, so they turned to their working-level colleagues in CIA for assistance. CIA military-economic analysts responded positively, and in effect CIA analyses were brokered by DIA to DoD consumers. This ad hoc arrangement was useful to analysts in both organizations: DIA was able to respond to its tasking, and CIA analysts were spared some of the presentational paperwork and other bureaucratic requirements that would have been necessary if requests had come directly to them. The informal arrangement worked well for a while, but the volume of support to DIA grew too large to handle this way. Secretary of Defense McNamara significantly raised the ante when in early 1963 he instructed DIA to include statements on economic impact and costs in all its military estimates. On May 29, 1963, McNamara's instruction was reported to Edward W. Proctor, then chief of CIA's Military-Economic Division in ORR, by a committee of senior DIA officials formed to survey the problem and make recommendations to the DIA director on how to implement the McNamara instruction.

This situation created a dilemma for CIA. On one hand, Proctor recognized from his conversations with DIA officers that they grossly underestimated the difficulties of developing a viable independent capability in this area. Worse, if they did manage to produce their own estimates of Soviet military spending, the existence of two sets of estimates would create chaos in the national security community and defense planners and policymakers would be hindered in their already difficult tasks. On the other hand, if DIA could not satisfy the legitimate needs of DoD consumers, CIA would have to fill the gap; Proctor simply did not have adequate analytic resources to meet his agency's and DIA's needs.

Proctor managed to work his way out of the bureaucratic trap. The memorandums for his superiors and for the record reporting on meetings, providing status reports, and recommending courses of action show that for the better part of the next two years he managed to provide assistance to DIA, at least for its highest-priority needs, thus keeping a difficult situation under control and ultimately maneuvering it to a workable solution. Finally, in early 1965, the working relationship between CIA and DIA in the area of military-economic analysis was firmly fixed in an exchange of letters between Director of Central Intelligence (DCI) John McCone and Deputy Secretary of Defense Cyrus Vance. The key aspects of Vance's response to McCone's proposal that responsibility for military-economic analysis be centralized were that "studies relating to cost and resource impact of foreign military and space programs . . . should be more centrally directed, monitored, and evaluated. I wholeheartedly concur that the Central Intelligence Agency should continue to have primary responsibility for these analyses; and, consequently, I support the expansion of the Central Intelligence Agency's capabilities in this area."[7]

Proctor's achievement served the national security community well for many years. From CIA's perspective the Vance letter accomplished two important objectives. By establishing CIA primacy in military-economic intelligence, it prevented an expensive, counterproductive duplication of effort. It also provided important ammunition for Proctor to use in the battle to increase CIA's military-economic analytic resources. Since DIA officers' first visit in May 1963, Proctor had pleaded with minimal success for additional analysts to help meet greatly increased demands within CIA and DIA. Bureaucracy being what it is, it took another year to get official authorization for a substantial increase in the number of military-economic analysts. At the same time, DIA was assured by the Vance letter that it would continue to receive support from CIA and have access to CIA's military-economic databases.

## Congress and the Spending Estimates

During the 1950s and 1960s congressional exposure to CIA's estimates of Soviet defense spending was limited mainly to what was included in annual briefings on the Soviet threat given by the agency's top executives in classified testimony before congressional committees. One exception was an article that Godaire prepared for Joint Economic Committee (JEC) hearings in 1962.[8] Godaire, who was chief of the Military-Economic Branch and worked under Proctor, discussed the possible coverage of the reported Soviet defense budget and possible defense-related budget residuals, as well as illustrative trends in total defense spending, outlays on military personnel, and procurement expenditures. He did not explicitly set out the agency's classified estimates.

The classified estimates did not come under regular congressional scrutiny until 1974, when the agency agreed to appear before Senator William Proxmire's Subcommittee on Priorities and Economy in Government of the JEC. DCI William Colby, in a prepared statement and in oral testimony, discussed the economic and military-economic situation in the USSR and China, in the process spending considerable time on CIA's estimates of Soviet defense program costs. The testimony was given first in executive session, but almost all of the proceedings were later sanitized and published by the U.S. Government Printing Office (GPO).[9] This procedure proved to be a template for hearings held almost yearly from then on under JEC auspices. CIA and DIA testified separately at first and then, beginning in 1986, jointly.

In opening the 1983 hearings Senator Proxmire noted that it was the tenth anniversary of the first hearing and declared that they had been "very valuable" in enhancing understanding of economic trends in the USSR and China and in setting out "in a candid and comprehensive manner the estimates and judgments of the Intelligence Community."[10] The senator and his staff assistant, Richard Kaufman, did their part to expand the dialogue by submitting lengthy lists of questions to CIA and DIA and publishing answers in the hearing records. These committee activities stimulated individual congressmen to request information on the defense spending estimates. Congressman Les Aspin was one of the interested customers. In 1975, for example, CIA answered Aspin's questions about expected growth in Soviet defense spending and the increase in the number of divisions in the North Atlantic Treaty Organizations (NATO) guidelines area.[11] Later that year Aspin followed up with a request for disaggregated information on defense spending for the years 1964–74. The agency was in the midst of a fundamental review of its estimates (see chapter 4), so its response was heavily caveated; it explained that since the last estimates were made in 1974, there had been changes in the manpower estimates and "new information on production levels and unit costs of Soviet weapons."[12]

Aspin's interest highlights one of CIA's concerns about releasing defense spending data outside the administration. As a critic of the Ford administration's defense policies, Aspin or any member of Congress could use them selectively or otherwise inappropriately in the debate over the defense budget. Suspicions of congressional motives came to a head when the Reagan administration had to deal with a Democratic Congress. Rather than let the JEC release its own version of the agency's annual testimony in the form of a press release (with the sanitized complete record following several months later), DDI Robert Gates ordered that the agency submission be unclassified from the beginning and given to the press at the time of the hearings.

### Arms Control and Military-Economic Analyses

With the exception of the Limited Test Ban Treaty signed in 1963, the arms control process did not gather much momentum in the policy arena until the Glassboro Summit Meeting in June 1967. Long before that, however, CIA's military-economic analysts were dealing on a regular basis with analysts in the economics division of the Arms Control and Disarmament Agency (ACDA). The economic dimensions of defense activity held a particular allure for disarmament program advocates. In the early 1960s many hoped that the complexities of intrusive physical inspections—which the Soviets consistently and adamantly opposed—could be obviated by economic measures to constrain defense activities. In its simplest form this consisted of nothing more than mutual limits on defense budgets. CIA military-economic analysts were frequently required to play the role of "skunks at the church picnic" in disabusing enthusiasts of the feasibility of such schemes. Over the years there were many variants on this general theme—some quite sophisticated—but in essence they all failed the test of practicality given the nature of the centrally controlled Soviet economy. Supporting this type of activity and ACDA's annual reviews of worldwide military expenditures was relatively inexpensive, but other forms of support to the arms control process were not.

The key policy question in this context was, "How strong are the economic motivations for Soviet interest in arms control?" This spawned hundreds of other questions that involved differing arms control and disarmament scenarios—for instance, "How much will the Soviets save if . . . ?" These kinds of questions chewed up enormous amounts of military-economic analytic resources and rightly so; the answers could play an important role in the formulation of U.S. national security policy. The following excerpt from an annual activity report of CIA's Military-Economics Branch illustrates the nature of support provided to the U.S. arms control process:

### Soviet Defense Spending

"Almost half of the branch effort in FY 70 was devoted to supporting SALT activities. Members of the branch served on Team 2 . . . and Team 4 . . . of the Arms Control Verification Working Group (an NSC body) and participated directly in the drafting of the *SALT Evaluation Report* for the NSC Staff in preparation for the Helsinki talks. Participation in and support to the SALT Backstopping Working Group included the drafting and coordination of SALT Contingency Paper 22: *The Soviet Military Budget as a Possible Monitoring Device*. The branch also participated in the two projects conducted by the NSC to update previous analyses and initiate new policy discussions in light of the experiences at Helsinki . . . and thus far at Vienna."[13]

## Organizational Responses to the New Demands

In the face of the explosion in new demands for the products of military-economic analysis, the logical organizational response would be an accompanying rapid increase in analytic resources to cope with these demands. A cursory look at the organizational evolution of ORR during the early 1960s suggests that this is in fact what happened. Until 1962 the Military-Economics Branch conducted the analysis of Soviet defense spending. In 1962 a military-economic division with six branches was created, and in 1964 the division was expanded further into two divisions and nine branches and was renamed the Military-Economic Research Area.

As is often the case, cursory examinations can be misleading. In fact, the number of people assigned to analysis of Soviet military spending was essentially the same after the two organizational changes as it was going into the 1960s—about a dozen people, including secretaries and research assistants. True, the new organizations reflected an increase in analysts, but they were all assigned to military analysis—particularly Soviet guided missile programs—and not military-economic analysis. (These were the days of the missile "gap" and the Cuban missile crisis.) Proctor, the primary force behind the expansion, had argued for the names Military Division and Military Research Area, but he was overruled by higher authorities, presumably in deference to the sensibilities of DIA and the intelligence components of the military services. The pre-1962 Military-Economics Branch was renamed the Military Expenditures Branch.

A review of Military Expenditures Branch activity reports and publications shows that demands were met quite another way. The branch ceased its research activities and focused almost exclusively on meeting demands for spending estimates and developing a computerized system to assist in the effort. We discuss the development of the computerized costing system in greater detail later, but it is worth noting that in the early 1960s the system was seen by military-economic analysts as a mixed blessing. It offered the only likely answer to escalating demands, but it was costly in terms of analyst talent diverted from research. By 1963, however, the system's payoff was clear to all.

Analytic resources' failure to increase in step with demand was not due to lack of effort. The following excerpt from a 1964 memorandum on the subject expresses Proctor's concerns: "This obligation to satisfy DIA-departmental requirements is only one of the additional demands for this type of work which have been levied on us. . . .We have taken on these tasks without any increase in the staff assigned to the problem. Even without the DIA-departmental requirements we would be hard pressed to handle these increased responsibilities with our present staff. The present situation is becoming intolerable. *We are currently meeting our obligations by drawing on past work which is becoming more and more obsolete and there are a number of critical areas of costing which need considerable new research*" (emphasis added).[14] Proctor's pleas were answered in 1966, when the Military Expenditures Branch split into two new branches—the Cost Analysis Branch and the Military-Economic Planning Branch—with a total staff of about twenty-five. Another major organizational shift occurred in July 1967 when CIA's Office of Strategic Research (OSR) was formed under the directorship of Bruce C. Clarke, Jr. OSR's creation was the belated institutional recognition of CIA's central role in military intelligence. There were two important consequences of this organizational change for the military-economic analysis function. On the positive side, Clarke, who inherited the two branches created in 1966 as part of the new OSR, shared Proctor's appreciation for military-economic intelligence and fully supported it. On the negative side, ambiguity developed about who was responsible for assessing the economic burden of Soviet defense activities. OSR was clearly responsible for estimating Soviet defense spending, and the new Office of Economic Research (OER), created at the same time as OSR from the remaining elements of ORR, was clearly responsible for estimating Soviet GNP. Analysts in the Military-Economic Planning Branch and the USSR Branch of OER had excellent working relationships, but these analysts marched to different managerial drummers whose concerns were somewhat different. As a result, some issues—particularly technical issues about how best to define and measure the burden of defense activity and the construction of dollar-to-ruble conversion ratios—fell through the cracks.

## The Products

The substantive focus of CIA's military-economic analytic products shifted markedly in response to the rapid escalation in demand in the early 1960s. In the 1950s and the early 1960s, although there were a few important exceptions,[15] most of the output consisted of published ORR papers documenting the database used in the costing process. The output of the costing process—that is, the expenditure estimates themselves—usually existed in detail only as typescript contributions to the pertinent NIEs, and they subsequently appeared in truncated form in the published NIE. In most of the 1960s and into the 1970s, the methodology papers nearly disappeared. Meanwhile the output of spending papers and memorandums proliferated, and many new channels were developed to get them to consumers.

Support to the NIE production process continued at a high level but became less central as direct lines of communication with policy consumers were established and used to convey estimated spending information, either in the form of typescript memorandums and oral briefings meeting specific requirements or CIA-published reports meeting the needs of a broader audience. Some of the formal products of the period are listed below.

CIA/DI, *SOVIET DEFENSE EXPENDITURES AND THEIR ECONOMIC IMPACT THROUGH 1970*, DECEMBER 1964. This was the first formal CIA publication on Soviet defense spending to reflect the impact of McNamara's Pentagon on intelligence support to the defense planning process. It presented the expenditure implications of the high and low alternative Soviet force levels developed by the intelligence community for the new national intelligence issuance, *Intelligence Assumptions for Planning, 1964*. In twenty pages of text and a fifteen-page statistical annex, the report presented high and low estimates of spending for 1961–70 in rubles and dollars on military mission and resource category bases.

CIA/ORR, *SOVIET DEFENSE EXPENDITURES, 1955–64, A CONTRIBUTION TO THE MEMORANDUM TO HOLDERS OF NIE 11-4-65, MAIN TRENDS IN SOVIET MILITARY POLICY*, JUNE 1965.

The existence of this paper reflects the burden and some of the confusion created by the proliferation of costing work. As the title hints, the costing work in support of NIE 11-4-65 was not completed in time to be included in the NIE, published in April 1965; hence, the requirement for a memorandum to holders of the NIE. Presumably the delay was caused by the groundbreaking work required for *Intelligence Assumptions for Planning* and other alternative force costing requirements, such as the work done for the CIA/DIA Joint Analysis Group. Almost half of the sixty-three-page CIA contribution covers a detailed explanation of differences in the spending estimates presented there and those presented for the previous NIE and for IAP-64. The number of different spending series appearing in the national security community at the same time was beginning to create some confusion. Indeed, the foreword notes, "The . . . section—explaining the differences between the current series and earlier series—has been included in response to several requests."

CIA/ORR, *ORR CONTRIBUTION TO NIE 11-4-66, MAIN TRENDS IN SOVIET MILITARY POLICY,* APRIL 1966. This CIA contribution to the NIE was unique; the time period the expenditure estimates cover extends back to 1950, reflecting a systematic ORR effort to create a historical database of forces and production information consistent with 1966 estimates of Soviet military forces and programs.

CIA/DI, *THE PRICE OF STRENGTH: BROADER SOVIET FORCE GOALS DRIVING UP DEFENSE SPENDING,* FEBRUARY 1968. Much of the analytic commentary included in the extensive CIA military-economic contributions to the NIEs of the mid-1960s ended up on the cutting-room floor of the interagency drafting and coordination process. After OSR's creation in July 1967, the scope of the CIA contributions was reduced and CIA turned to its own departmental publications as a more effective formal way to disseminate its analytic judgments. This paper was the first example of the new approach. It focused on the Soviet context by presenting an interpretation of historical and current ruble spending patterns linked to key decision points in the evaluation of Soviet defense policy. No dollar comparisons with U.S. spending were included.

CIA/DI, *THE 1969 SOVIET DEFENSE BUDGET,* JANUARY 1969. Although earlier CIA reports had presented analyses of the overall Soviet budget, this was the first of four annual reports that used the USSR's official announcement of its defense spending to present the CIA estimates.[16] These annual reports grew in size and complexity from thirteen pages in 1969 to forty-six pages in 1972.

CIA/DI, *SOVIET SPENDING FOR DEFENSE: AN ANNUAL REVIEW, VOL. 1, TRENDS IN RUBLE EXPENDITURES*, AND *VOL. 2, A MONETARY COMPARISON OF SOVIET AND U.S. DEFENSE ACTIVITY*, AUGUST 1973. A new presentational approach was initiated in 1973. The estimates were published in two separate volumes—one covering ruble data and the other covering dollar valuations of Soviet programs—totaling about 120 pages. The ruble paper presented the data "from the perspective of Soviet defense and economic planners" so that information was structured "in a manner believed conceptually consistent with Soviet defense accounting and economic statistical practices." The dollar paper went into considerable detail explaining the conceptual and methodological difficulties of constructing a useful, legitimate comparison framework.

Based on the volume of requests for follow-up information and analyses generated by these and other formal publications, there is little doubt that CIA's military-economic analyses had a substantial impact on national security planning and policymaking. Nonetheless, we sense that the most significant impact occurred in response to specific requests in the form of less formal products. Of course, many requests never would have been made, nor fruitful lines of communication established, without the appearance of the formal publications.

## Some Unhappy Customers

It is not surprising that, as CIA's military costing work was brought to bear in an analytic context, as in the military planning process in the Pentagon, or in a political context, as in the annual congressional debate over the U.S. defense budget, its validity was questioned by those whose oxes were likely to be gored with the help of the CIA data. For example Admiral Elmo Zumwalt, Chief of Naval Operations in the early 1970s, was dismayed by CIA estimates of Soviet spending for naval programs. The estimated dollar values of Soviet programs were generally lower than U.S. spending, and this did not square with the admiral's perceptions and obviously was not helpful in the budgetary debate. Zumwalt made his concerns known to the agency, and one of us was dispatched to provide detailed briefings on CIA naval spending estimates and their underlying methodology, first to senior analysts at the Center for Naval Analysis and then to Zumwalt and a few senior officers from his systems analysis group, OP-96. Zumwalt was polite, but he was clearly not persuaded that CIA knew what it was doing in its costing work. Clarke, director of OSR, invited Zumwalt to assign an officer of his choosing to take up residence with CIA's military-economic analysts for as long as necessary to familiarize himself with the data and procedures used to cost the Soviet navy. The invitation was accepted, and Commander Nicholas Brown spent several weeks at CIA. Brown made some minor suggestions for improvement which were incorporated by OSR analysts, but he concluded that overall the CIA work was valid and commented favorably on it.

A more serious complaint was raised in March 1972 at a meeting of the United States Intelligence Board (USIB) when Lt. Gen. Donald Bennett, Director of DIA, and Brig. Gen. Daniel Graham, Director of Estimates at DIA, challenged the CIA expenditure estimates' validity and the appropriateness of their use in the NIEs. This complaint undermined the sensible and cordial working arrangement between CIA and DIA which ensured that the agencies would be working from the same set of books. If DIA's leadership was now rejecting the validity of CIA work, it seemed inevitable that it would begin producing its own estimates, and chaos (and needless extra cost) would ensue. Another potentially serious problem with the Bennett/Graham complaint was that it included a proposal to convene a conference of nongovernment experts to discuss the general subject of estimating Soviet military spending. CIA viewed this proposal with dismay.

## Creation of the Military-Economic Advisory Panel

Proctor, who had risen to the position of CIA's DDI in the early 1970s, worried that the DIA attempt to enlist a large number of outside experts and contractors with no or low levels of security clearance would only result in confusion.[17] He also sensed that DCI Richard Helms was increasingly uneasy about the spending estimates. In order to reassure Helms, head off the potentially disastrous effects of the Bennett/Graham complaint, and clear the air, Proctor wrote to Helms requesting that the DCI direct the DDI to "establish a small panel of recognized specialists outside of government who can examine carefully our methods and our data, and do it on an all-source classification level."[18] On the routing slip to this memorandum, Proctor further pointed out to Helms that setting up the panel under DCI sponsorship could "discourage DIA from continuing with its unsubstantiated badmouthing of existing estimates" and "provide the entire community with an unbiased assessment of our estimates and how these might be improved."

Helms approved Proctor's recommendation the day he received it and informed Bennett of his decision.[19] Almost a year passed, however, before members of what was called the Military-Economic Advisory Panel (MEAP) could be recruited and fully cleared for access to the information available to the military-economic analysts. MEAP met for the first time April 6–9, 1973, under the chairmanship of Professor Holland Hunter of Haverford College, a widely recognized expert on the Soviet economy.[20] After three more meetings, the panel issued its first report in July 1974.[21] In its report MEAP defined its purpose as trying to help CIA deal effectively with what it termed the "military-economic intelligence quandary." As MEAP viewed it, "The U.S. national interest requires careful evaluation of Communist military and economic activity. Its dimensions and details are complex and very incompletely revealed by the countries involved. Serious differences of opinion face U.S. policymakers in evaluating available evidence. The problem is to minimize uncertainty and inconsistency, and to marshal the evidence persuasively in forms directly applicable to decisionmaking."

MEAP endorsed CIA's analytic effort on aggregate money values as a "common denominator" for the analyses of changes in Soviet defense programs, urging continuation of direct costing of defense as well as estimates of annual changes in Soviet GNP. But the panel noted that the complexities of comparing defense efforts in different countries had led to confusion among potential users of CIA's spending estimates. It found that some were "highly skeptical of the estimates" and others thought detailed costing should be given up altogether.[22] And it concluded that efforts to explain the concepts and methodology behind the estimates had been insufficient and recommended that OSR prepare a methodological manual to explain the costing process to new analysts, analysts in other agencies, and "skeptics all around town."[23] (Later in this chapter we discuss additional recommendations for improvement in CIA's military-economic analyses made in MEAP's first report.)

MEAP remained an active advisory group for the next twenty years. Although its focus shifted depending on the incumbent DCI and DDI, it continued to provide an unbiased outside perspective on the CIA military-economic effort. Former members told us that MEAP required far more work on the part of the panel than is usually the case in such consultative bodies and they thought the efforts had a constructive impact, particularly in influencing the content and priorities of CIA's military-economic research activities.

## Changes in Methods and Evidence

### Automating the Costing Process

Even before the Kennedy administration increased demands for military-economic analyses, it was clear to CIA/ORR managers and analysts that new techniques in data handling were required to take full advantage of the building-block methodology. Having been the principal drafters, we can attest that hundreds of hours of analyst time were spent on calculations and data presentation in the production of *Soviet Military Expenditures, 1955–61* and *Estimated Soviet Military Spending, 1955–1961, by Military Mission* in late 1960 and early 1961. These two reports were a giant step forward in the presentation of detailed spending data, but they still only scratched the surface of what was possible. The analyst time devoted to the mechanical aspects of the approach would have been far better spent on research.

The mantra of cost-benefit analysis pervading the Pentagon after the arrival of McNamara and RAND magnified the challenge confronting CIA's military-economic analysts. It also confirmed the basic wisdom of those who chose the building-block approach over the macro approach to estimate Soviet defense spending in the mid-1950s. Questions of this type were posed: "Which of the following alternative means of delivering $z$ megatons of nuclear warheads to targets $a$ and $b$ would be the lowest cost choice for the USSR: ICBM system $x$, SLBM system $y$, or bomber system $z$?" Questions like this bear little resemblance to their progenitor: "How much is the USSR spending for defense?" Clearly, analysis of Soviet economic statistics would not have helped U.S. military program planners of the 1960s.

Fortuitously, CIA was in the early stages of productively applying automatic data processing (ADP) to intelligence problems. The DI created a small ADP applications staff as part of this effort, and in late 1960 the staff conducted a survey of the directorate to identify possible computer applications. It concluded that ORR's analytic effort on Soviet defense spending constituted an area of intelligence production that would be greatly aided by an appropriate application of ADP techniques. In early 1961 the concept for a computer-based military-economic system was developed; the system would store pertinent data on quantities and prices and would perform all calculations required to produce appropriately categorized expenditure estimates. The system was named the Strategic Cost Analysis Model (SCAM).[24] Not long after, a small team of systems analysts and programmers from the agency's Office of Computer Services (OCS) and military-economic analysts from ORR was assigned the task of jointly developing the system.

At the time progress developing the system seemed torturously slow. Two years passed before the computer system was generally functioning to its design specifications. Given the size and complexity of the system—it used more central processing unit time than most other CIA computer applications processing on mainframe computers—and the fact that a major delay in progress was caused by nonconcurrent moves of the military-economic analysts and the computer services component from downtown Washington to Langley, Virginia, the SCAM system was retrospectively viewed as a major success in achieving the design criteria originally established.

The typical experience in the computer field was that the development of new computer application systems on a comparable scale took longer or never made it to operational capability. The successful development of the military-economic computer system can probably be attributed to two factors: demand and people. The demand for analytically sophisticated military-economic intelligence products was incessant and created pressure that could only be relieved with the aid of a computerized system. Fortunately managers in ORR and OCS recognized that success was only likely if good people were assigned to the joint developmental effort and, perhaps most important, if they were allowed to remain with the project from beginning to end.

The SCAM system in the early 1960s consisted of three main data files.

THE MANPOWER/ORDER-OF-BATTLE FILE. This file contained more than six hundred distinct types of military units, a personnel manning factor for each type, and the number of units in the Soviet order-of-battle for each year of a thirty-year span. The units (e.g., Bison bomber regiments) were grouped into thirty functional elements (e.g., intercontinental attack forces) and ten functional missions (e.g., strategic attack forces).

THE EXPENDITURES ACCOUNT FILE. This file consisted of forty expenditure accounts that could be directly linked to the units defined for the manpower/order-of-battle file (e.g., the annual pay and allowance bill for a Bison bomber regiment). On average, ten unit-related expenditure factors in this file would be associated with a given type of unit.

THE PRODUCTION/PROCUREMENT FILE. This file contained information on the estimated numbers produced and the estimated ruble and dollar prices of more than five hundred items of military hardware. Combining the production data in this file with the price information produced procurement expenditure estimates that were then allocated to the appropriate units in the order-of-battle file.

Utilizing the system involved revising any of the information in the basic data files and generating new output information. The output formats available included a manpower/order-of-battle report; two production/procurement reports, one with procurement expenditures in rubles and one in dollars; two estimated expenditure reports summarizing expenditures by major military missions and functional elements, one in rubles and one in dollars; and two estimated expenditure reports that included spending at the individual unit level in addition to the functional mission and element summaries, one in rubles and one in dollars. All output information could cover a thirty-year period. A complete set of output listings created about a thirty-inch stack of 14" x 20" computer paper.

One important measure of SCAM's success is the fact that it endured for thirty years as the backbone of CIA's military-economic effort and was still going strong when the detailed work ceased at the Cold War's end. The system was continually revised to accommodate advanced computer hardware and software and refined and expanded to reflect advances in costing methodologies, but the essence of the system remained the same. The obvious benefit of the SCAM system was that it permitted a response to almost any question about Soviet defense spending. Not so obvious but equally important benefits included the disciplined framework that the basic data files provided for storing and retrieving key intelligence estimates of the physical dimensions of the Soviet military establishment, and the research management tool it provided by highlighting weak points in CIA's overall understanding of Soviet military forces and programs. These benefits, of course, did not come without cost. Automating the process did not reduce the number of people doing the job, but it did make possible a greater volume and variety of output produced more efficiently and in greater detail. Indeed, although it is difficult to measure with any precision, the resources devoted to the care and feeding of SCAM probably grew from the mid-1960s to the mid-1970s. The most significant cost of SCAM, however, was its opportunity cost: the basic research on Soviet military prices and cost factors that was seriously delayed or never done.

## Better Evidence on Military Forces and Programs

On August 18, 1960, the first operational U.S. photoreconnaissance satellite system acquired its first target in the USSR—Mys Schmidta, a Soviet long-range air forces staging base airfield 425 nautical miles northwest of Nome, Alaska. Although the advent of satellite imagery systems did not answer all of the intelligence questions about the thousands of quantitative and qualitative measures in the Soviet military establishment, it did usher in a prolonged period of rapid improvement in the scope, depth, and quality of intelligence information on Soviet military forces and programs.

By the mid-1960s the nature of the disagreements among military analysts had changed drastically. Before the availability of pertinent imagery, the argument might have been over whether twenty or thirty submarines of a given class had been produced, but now the argument was more likely to be over whether the submarine launched last Thursday was the twenty-third or twenty-fourth in the series, and whether it was launched Thursday morning or Wednesday afternoon. The new information not only made it possible to count and measure things more reliably—particularly large, costly items like ships, submarines, and missile sites—but also to monitor changes more accurately. Given the direct link between the underlying physical base and the spending estimates implicit in the building-block method, these improvements in the quality, reliability, and timeliness of evidence had a profound impact on the confidence of those producing the estimates.

The military-economic analysts, of course, were not alone. Without the larger intelligence community's increased confidence in its improved understanding of the Soviet military establishment and its ability to closely monitor it, the U.S. government would not have had the confidence to pursue a policy of arms control with the USSR. A constructive, mutually reinforcing cause-and-effect relationship was thus established. Better intelligence meant greater policymaker interest in arms control, which resulted in greater pressure and increased resources to produce more improvements in intelligence, which further raised interest in arms control policy, and on it went. An important byproduct of the process, of course, was an improved physical base for the spending estimates.

## Other Improvements in Costing

Important progress also was made during the 1960s on the dollar prices and cost factors used in the costing process by obtaining more detailed and reliable information from the Pentagon and through contacts with U.S. weapons manufacturers. In addition, there were the improvements in costing techniques and the presentation of results made possible by SCAM. Although basically mechanical in nature, many of these improvements enhanced the quality and utility of the spending estimates. Perhaps the best example is provided by the system developed to compute directly from order-of-battle data the requirements for nuclear materials, production of specific warheads, procurement costs of the material and warheads, and the allocation of the procurement costs by military mission and element. Comparable systems were developed for other classes of procurement.

## Transition to 1967 Ruble Price Base

The USSR carried out a reform of wholesale prices in 1967, hoping to achieve a better relation between costs and prices and to eliminate most loss-producing enterprises.[25] The 1955 prices no longer reflected the costs of production in the late 1960s. OSR's 1955 price base for its defense spending estimates was out of date for the same reason. Indeed, some Soviet sources indicated that part of the rationale for the 1967 reform was to bring prices for military goods in line with civilian product prices.[26] In the spring of 1969, OSR analysts worked out a procedure for converting 1955 prices to 1967 prices and told their superiors that "base conversion is a step that has been delayed too long."[27]

The problem facing analysts, however, was that although they had information on the extent of price changes for goods like food, clothing, electricity, and petroleum, oil, and lubricants (POL), they did not know how prices on defense hardware had changed. So they used the published information on price increases by industrial sector. The defense purchasers of goods and services represented in SCAM were sorted out by the sectors in a western reconstruction of a Soviet input-output table. These values in the table were then multiplied by sector price conversion factors (1967 prices divided by 1955 prices) derived primarily from Soviet publications. Where conversion factors could not be based on open sources, they were estimated by examining the composition of inputs into the manufacture of defense goods (steel, energy, labor) and calculating a weighted price index from price changes reported for individual inputs. After the worry about how to convert to a new price base, the end result was anticlimactic. Table 3.1 shows the defense spending estimates for 1968 expressed in 1955 and 1967 prices.

Of the 9 percent increase in the estimate, two-thirds could be attributed to the price reform and the rest to increases in food prices and wages of civilian personnel during 1956–67. Within the procurement category, prices for land arms were raised by 25 percent in the conversion, and prices for naval ships and boats increased by 12 percent. Prices for electronic equipment were unchanged. When OER put its Soviet GNP accounts on a 1970 price base in the early 1970s, defense spending estimates were also treated as being in 1970 prices, on the assumption that changes in Soviet prices between 1967 and 1970 were inconsequential.

In effect, the move to a 1967 price base for defense goods and services began with a 1955 ruble price base that rested on a slender foundation of actual Soviet prices with respect to procurement and much of operations and maintenance (O&M). It then accepted Soviet price indexes as a valid measure of price change between 1955 and 1967 for both civilian and defense goods. Finally, the conversion accepted Soviet announcements of price change in the 1967 reform as applicable to defense goods. When ruble prices for Soviet military equipment in the 1960s and early 1970s began to accumulate in the agency's files, these assumptions proved unfounded; the estimated 1970 prices for defense hardware in the form of original equipment and spare parts were badly understated.

TABLE 3.1
SOVIET DEFENSE SPENDING IN 1968 IN 1955 PRICES AND 1967 PRICES

|  | Billion 1955 rubles | Billion 1967 rubles | Ratio '67 rubles to '55 rubles |
|---|---|---|---|
| RDT&E | 6.11 | 6.29 | 1.03 |
| Procurement | 5.62 | 6.20 | 1.10 |
| Construction | .40 | .47 | 1.18 |
| Personnel | 5.08 | 5.17 | 1.02 |
| O&M | 3.70 | 4.76 | 1.29 |
| Total[a] | 20.91 | 22.89 | 1.09 |

[a]Totals are based on unrounded data and may not equal the sum of the rounded components shown in the table.

Questions, Techniques, Evidence

Near the end of the period, in the early 1970s, some military-economic analysts began to have serious doubts about the ruble price base used for estimating procurement expenditures. It had been too long since the price base had been systematically reviewed, and the levels of estimated total ruble spending for military procurement just seemed too low. Moreover, the bits of relevant ruble price information were difficult to reconcile with the procurement ruble price base in use by the SCAM system.[28] An initially small but focused effort was launched to systematically review the ruble prices. We discuss this effort in chapter 4.

## Areas for Improvement

MEAP RECOMMENDATIONS. In its initial report, MEAP emphasized a number of themes, to which it returned in subsequent reports. These included the desirability of estimating equipment inventories in value terms, increased research on the stages of the RDT&E and production cycles, the need to explain to users reasons for year-to-year changes in the estimates, the need to provide some sense of the uncertainty surrounding the estimates, the additional perspective to be gained by interindustry analysis of the impact of defense spending on the economy, and more analysis of investment constraints on the rate at which large defense programs could be implemented. These recommendations constituted an ambitious strategic plan that had a major impact on shaping the research programs for military-economic analysis in years that followed.

CONTROVERSY OVER THE DEFENSE BURDEN. The weaknesses of the ruble price base became apparent in retrospect when Secretary of Defense James Schlesinger asked in 1974 for a paper on the USSR's defense burden. While Schlesinger had been DCI, he had thought the existing estimate of the ratio of defense spending to Soviet GNP was out of line: "But the 6 or 7 percent at the time I came to the Agency just struck me as wrong. Because you looked at this society and the goddamned society was militarized and you knew the society had not become militarized with such a small fraction—Hell, we were spending 6 percent of the GNP or thereabouts at the time."[29] The paper CIA prepared, however, concluded that the burden was probably about 6 percent of GNP in 1972 and less than 8 percent even when some possible additional costs were included.[30]

When the DDI, Ed Proctor, sent the paper to Schlesinger with the notation that he intended to release an unclassified version, Graham, then DIA director, protested vehemently. Graham stressed the difficulty in comparing market and nonmarket economies, the distortion in Soviet pricing, the primacy of national power in leadership thinking, and the high share of certain critical technologies monopolized by the Soviet military. But his principal worry seems to have been the "possibility, perhaps probability, that non-economists will reason from such comparisons that the Soviet defense effort is substantially smaller than our own, because it absorbs a roughly equal percentage of an economy about one-half the size of ours (as measured in rubles converted to dollars)."[31] Proctor thanked Graham for "thoughtful and well-reasoned comments" and explained that the paper did not intend to deal with Soviet political or power aims or the scale of the USSR's defense programs in dollar-equivalent terms. Rather, it aimed to analyze the burden from an economic standpoint and leadership perceptions. The DDI promised, however, to revise the paper to eliminate some possible sources of misinterpretation and to issue it on an "official use only" basis to a small number of people working in the Soviet field.[32]

The burden paper was finally published in April 1975.[33] Its "most important conclusions" were that for Soviet leaders costs seem not to "have been a major factor in their military decisions" and "are unlikely to constrain the Soviets unduly in the future." It provided various measures of the burden (less than 8 percent of GNP, less than 10 percent of the labor force, 10–15 percent of industrial production, and 20–30 percent of machinery output). The paper also included a discussion of the "GNP paradox" to deal with Graham's concern about noneconomists' tendency to confuse burden measures with size measures. To this same end, the paper presented a dollar comparison of U.S. and Soviet programs, saying Soviet programs "have exceeded U.S. expenditures each year since 1971" and were 20 percent higher in 1974. Finally the paper included an input-output analysis of the shares of output in the so-called material sectors of production going directly or indirectly to the Soviet defense effort. It was the most exhaustive treatment of economic burden and leadership perceptions of burden undertaken until that time. Unfortunately the statistical basis for the macro measures of burden (share of GNP, industrial output) would be demolished within the year when OSR introduced much higher estimates of procurement of weapons and spare parts, as well as RDT&E. The problem was not distortion in Soviet pricing or differences between market and planned economies but CIA's knowledge of prices for Soviet defense hardware.

Questions, Techniques, Evidence

RESEARCH ON RUBLE PRICES. It is fair to say that research on ruble military prices—particularly the prices of military hardware—was generally neglected from the early 1960s to the early 1970s. As we have noted, an effort was made to obtain improved information on dollar-equivalent costs, but work on ruble prices until the last few years of the period was essentially passive. The few pieces of pertinent information that became available were reviewed and filed but not deeply analyzed; individual analysts were able to concentrate on specific ruble price issues only sporadically.[34]

The neglect of ruble prices of military hardware resulted from a combination of factors—lack of adequate resources, priority demands for spending estimates, and some misplaced confidence. The military-economic analysts justifiably had high confidence in their understanding of Soviet military personnel costs in rubles. The research base was solid and periodically updated with an abundant supply of new information. Unfortunately this confidence tended to spill over into the area of military equipment prices where it was not justified. The research base on military equipment prices and dollar-to-ruble conversion factors created in the 1950s was good, but it was not as extensive or solid as the personnel cost base. And it had not been periodically updated in any systematic way.

Given the explosion in demand for Soviet military spending data; the immediate demands on analyst time caused by the design, development, and maintenance of the SCAM system; and a limited supply of analytic resources, it is not surprising that the difficult research task of reviewing and updating the ruble price database was repeatedly delayed. The cracks in this crucial part of the overall foundation were perceived, but it wasn't until the small but concentrated research and collection effort in the early 1970s that this perception became clear. By then it was too late to prevent the collapse of an important part of the costing edifice. We discuss the high price of this collapse, in terms of lost confidence in CIA's military-economic work and the agency's efforts to rebuild, in the next chapter.

# Chapter 4
# The Estimates, 1975-90
## Substantial Progress and Managerial Doubts

The most important event in the history of CIA's military-economic analysis of the USSR was the early decision to rely principally on the building-block methodology. This decision determined the nature and size of the analytic effort required to produce spending estimates, as well as the utility and impact of the estimates in the policy community. In retrospect, it is clear that the second most important defining event was the major upward revision in the ruble estimate of Soviet military spending announced by CIA in early 1976. This event had a profound, lasting impact on the evolution of CIA's work on the subject.

One major consequence of the 1976 revision was the focus it created within CIA on the need for a substantial increase in analytic resources to restore and maintain the basic knowledge necessary to produce accurate spending estimates. Such an increase and its desired payoff did, in fact, occur. The improvements over the next fifteen years in the depth and breadth of pertinent knowledge and in the sophistication of estimating techniques is, we believe, impressive. These improvements provided a solid base for increased confidence in the work.

From one perspective, the upward revision might have been expected to increase confidence in the CIA work. After all, those responsible for producing the revised estimate were demonstrating that they were more concerned with using available evidence to provide the best possible estimates than they were about criticisms a major upward revision would incite. The reaction to the revision was quite different. The shock of the abrupt change created deep and lasting skepticism among many in the policy community, the senior management ranks of the intelligence community, and the academic community. The central irony of the history of CIA's military-economic analysis was that institutional support for the analysis and its acceptance in the policy community was at its weakest when the work itself was at its best. From just about any point of view, the 1976 revision can be characterized as a bombshell.

The last fifteen years of CIA's program to estimate Soviet defense outlays were marked by other significant changes in estimates, methodologies, and organizational structure, and by the activities of external review panels. These events and activities combined with the 1976 revision to create the most tumultuous period in the history of CIA military-economic analysis. The key events and their timing are depicted in figure 4.1. We begin with a description of the circumstances surrounding the abrupt increase in the ruble estimate in 1976, the furor it produced, and its effect on the credibility of the agency's costing effort. When the specific new intelligence that served as a catalyst for the revision was uncovered, OSR was already well on its way to making substantial improvements in the dollar procurement cost estimates and had developed growing doubts about the ruble procurement prices it was using. It was also in the process of incorporating new evidence on Soviet operating practices and RDT&E programs in both dollars and rubles. In reaction to the important new evidence and its growing unease, the agency mounted a campaign to obtain enough ruble prices to provide a stronger base for its ruble procurement estimate.

In the early 1980s, in the midst of the major research generated by the

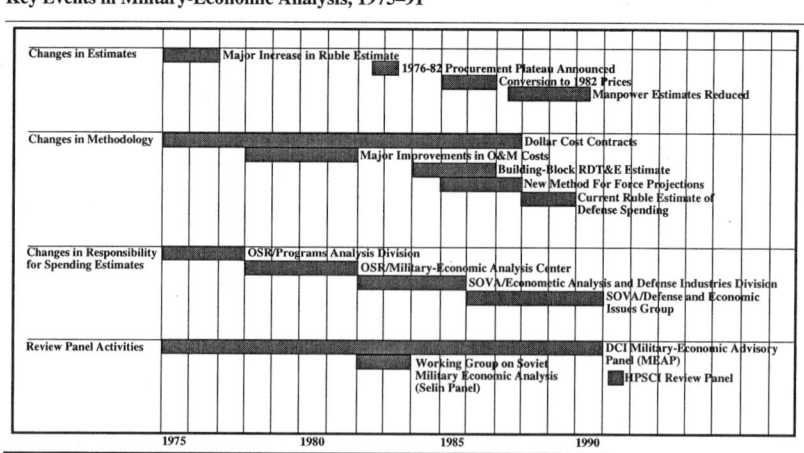

Figure 4.1
Key Events in Military-Economic Analysis, 1975–91

controversy surrounding the 1976 revision, CIA concluded that real growth of Soviet defense spending had slowed in the mid-1970s, contradicting its earlier report of sustained rapid growth since the early 1960s. Coming at a time when the U.S. defense budget was climbing steeply, the CIA revelation of a slowdown in Soviet spending contributed to skepticism directed at the spending estimates. This chapter concludes with a review of the factors that led CIA's senior management to back away from the estimates and reduce the costing program to a mere shadow of its former self.

## The Bombshell

In May 1976 CIA published the findings of a major reassessment of Soviet ruble defense spending in classified and unclassified papers.[1] Intended as interim reports on the impact on the ruble estimates of "an unusually large body of new information," the papers said:

- *Soviet defense spending (as the USSR might define it) had increased from 40–50 billion rubles in 1970 to 55–60 billion rubles in 1975, measured in constant 1970 prices. (A year earlier the comparable figures were 29 billion rubles for 1970 and 34 billion rubles for 1975—an increase of 38 to 72 percent for 1970 and 62 to 76 percent for 1975.)*
- *Over the same period the rate of growth of defense spending in 1970 prices was now thought to be 4 to 5 percent per year instead of the 3 percent per year reported in previous estimates.*
- *Since 1970, defense requirements had been absorbing 11–13 percent of GNP (vice less than 8 percent), depending on whether defense was defined narrowly (U.S. definition) or broadly (Soviet definition).*

At the same time, the papers emphasized that because about 90 percent of the increase in ruble estimates resulted from "changes in our understanding of ruble prices and costs" rather than "discovery of larger programs," the revision did not affect CIA's appraisal of the size or capabilities of Soviet military forces or change appreciably the estimated dollar cost of reproducing Soviet defense programs.

After the sifting of evidence was completed in 1977, the changes in the ruble and dollar estimates were as shown in table 4.1. The 1977 estimate of total cumulative Soviet defense spending in 1971–75 (in 1970 rubles) increased 54 percent over the 1975 estimate for the same period. Almost all of the rise was the effect of higher ruble-to-dollar ratios used for estimating

## TABLE 4.1
## CIA'S ESTIMATES OF CUMULATIVE DEFENSE SPENDING, 1971-75: PRE- AND POSTREVISION

|  | Best-75[a] | Best-77[b] | Absolute Increase | Percentage Increase | Best-75[c] | Best-77[d] | Absolute Increase | Percentage Increase |
|---|---|---|---|---|---|---|---|---|
|  | (Billion 1970 Rubles) | | | | (Billion Dollars) | | | |
| Total defense[e] | 151.55 | 233.78 | 82.23 | 54.3 | 503.67 | 556.51 | 52.84 | 10.5 |
| Personnel | 33.91 | 37.81 | 3.90 | 11.5 | 237.63 | 217.60 | -20.03 | -8.4 |
| Procurement | 56.39 | 94.04 | 37.65 | 66.8 | 107.08 | 130.21 | 23.13 | 21.6 |
| O&M | 30.71 | 49.18 | 18.47 | 60.1 | 96.08 | 127.82 | 31.74 | 33.0 |
| Construction | 2.52 | 6.21 | 3.69 | 146.4 | 6.84 | 13.59 | 6.75 | 98.7 |
| RDT&E | 28.04 | 46.54 | 18.5 | 66.0 | 56.06 | 67.28 | 11.22 | 20.0 |

[a]OSR Best-75, September 3, 1975. The label "best" was used to identify spending estimates reflecting the prevailing intelligence judgments on the most likely levels of underlying Soviet military forces and programs, and to distinguish them from the many "what if" excursion spending estimates made for a variety of analytic purposes.

[b]OSR Best-77, September 7, 1977. This estimate contains some outlay categories that were not included in the 1975 estimate: within procurement, aircraft ground support equipment and vehicles (.71 billion rubles); within personnel, military pensions, Ministry of Defense social insurance payments, and medical and clothing costs (4.30 billion rubles); and within O&M, pensions and social insurance payments for civilian employees of the Ministry of Defense and costs associated with preinduction military training (1.59 billion rubles).

[c]Best-75 is compiled in 1974 dollars, and Best-77 is in 1975 dollars.

[d]OSR Best-77, September 7, 1977. Like the Best-77 ruble estimate, this estimate includes some outlay categories not found in Best-75: in procurement, this estimate includes some aircraft ground support and vehicles ($1.33 billion); in personnel, medical costs and pensions and social security payments for military personnel ($11.20 billion); and within O&M, preinduction military training ($11.51 billion).

[e]Totals are based on unrounded data and may not equal the sum of the rounded components shown in the table.

the procurement of weapons and equipment, the completion of a research effort that exploited a Soviet costing model for ships and boats, the rise in estimated maintenance costs for weapons and equipment in parallel with the increase in their estimated acquisition costs, and a revised estimate of military RDT&E.

Unlike the ruble spending estimates, the dollar-equivalent costs of Soviet defense programs in 1971–75 did not change significantly in real terms in the 1977 revision. More than half of the 10 percent increment in estimated dollar costs shown in table 4.1 can be attributed to the difference between the 1974 prices used in Best-75 and the 1975 prices embodied in Best-77.[2] The Best-77 estimate also included outlays for some activities that had not been covered in earlier estimates. At the same time, personnel costs were pulled down by a decision to assign U.S. pay rates to Soviet servicemen according to the jobs they performed rather than the rank they held. In addition, the difference in the dollar costs between the two estimates was reduced by the correction of an error in the calculation of the Best-75 estimate of the dollar cost of Soviet RDT&E.

To this day, there is disagreement about the roots of the 1976 revision. Some critics outside CIA have said it was caused by new intelligence acquired from an emigré in 1975. Donald Burton, head of the CIA division responsible for defense costing, has said that the 1975 intelligence was an important catalyst, but in the early 1970s a body of evidence accumulated indicating that the 1970 ruble prices used for procurement in SCAM were too low. During this period, CIA had also gained possession of a ship costing model used in the USSR, which—when given the physical parameters of some Soviet naval ships—turned out prices that fit well with the few prices recently collected, and indicated ruble-to-dollar ratios for naval ships three times higher than the ones that had been used. Meanwhile, new information on the relation of total Soviet outlays on RDT&E to officially reported outlays on "science" had caused a large upward revision in ruble estimates of military RDT&E.[3] OSR's Programs Analysis Division planned to introduce the higher ruble-to-dollar ratios gradually as new prices were analyzed and accepted, but the volume of new prices argued against gradualism.[4]

In late 1974 CIA received a report on a Soviet economist who had recently emigrated legally to the West. Far back in the report the interviewer noted that the emigré claimed knowledge of the Soviet defense budget. Additional questions were submitted to the interviewer by CIA and DIA analysts, and on the basis of the answers a CIA/DIA team was sent in April 1975 to debrief the emigré. Following the team's report, the emigré was brought to Washington for follow-up questioning in June–July 1975. The emigré claimed that while he was employed on a project to forecast consumer demand in the USSR for the 1971–75 state economic plan, he visited the USSR Central Statistical Administration, where he saw a document titled "Estimate of Expenditures for the Financing of Non-Productive Goals." The document reported actual outlays for 1969 and preliminary

1970 data in several sectors that the Soviets classified as "nonproductive"—education, health, defense, and others. The outlays were disaggregated in categories equivalent to those used in official reporting on expenditures by state-supported institutions in the USSR's state budget. Within the overall defense category, the emigré focused mainly on outlays for wages and quartermaster supplies in connection with his assigned project, but he also remembered the totals allocated to defense and later filled in the other entries in table 4.2. A task force of military-economic analysts from OSR and economic analysts from OER quickly formed to assess the emigré's information. In the midst of this review, the DDI met with analysts from OSR and OER to impress on them the importance of obtaining "a new appreciation of Soviet military expenditures based on this new information" as soon as possible.[5] The emigré's information strengthened the agency's confidence in the higher ruble prices it was acquiring and led it to accelerate its introduction of new ruble-to-dollar ratios for procurement.[6]

For years thereafter, contradictory charges were made (sometimes by the same people) that the agency either balked at accepting the emigré's information because it showed higher levels of defense spending than CIA's estimates, or that it adopted the information wholesale as a means of escaping an analytical cul-de-sac. As we indicated in chapter 3, CIA's estimate that defense accounted for only 6 percent of Soviet GNP had already inspired doubts about the ruble spending figures.[7] Accusations of skullduggery were leveled about the handling of the emigré's information. The Committee on the Present Danger claimed that CIA "reportedly attempted to discredit the source with an improperly administered polygraph."[8] This was untrue.[9] Others thought CIA had tinkered with the ruble-to-dollar ratios to align the agency's procurement estimate with that given by the emigré. Steven Rosefielde, observing that the revision in the estimates had no effect on real procurement growth, wrote that "CIA very cleverly and effectively continued to bury the contradictions of the past."[10]

What really happened? The comparison of the revised CIA estimates for 1970 with the emigré's reconstruction of what he said he had seen as a preliminary breakdown of defense outlays in 1970 shows that the revision did not simply replicate the emigré's account. The emigré's figure for planned outlays for defense in 1970 was 49.6 billion rubles, whereas CIA's revised estimate for that year was 40.6 billion rubles. Nor was there any mechanical raising of the ruble-to-dollar ratios.[11] The fact is that the collection and analysis of new prices in the early 1970s demonstrated that the initial conversion from 1955 ruble to 1970 ruble prices had been badly flawed.

## TABLE 4.2
## THE EMIGRÉ'S RECONSTRUCTION OF
## "EXPENDITURES FOR DEFENSE" SECTION OF THE OTSENKA[a]
(billion rubles)

| No. | Title of articles | Previous reporting period, 1969 | | Reporting period, 1970 (expected fulfillment) | |
|---|---|---|---|---|---|
| | | Amount | Percentage of Total | Amount | Percentage of Total |
| 1. | Expenditures for the increase of fixed capital | 20 | 42.1 | | |
| 2. | Expenditures for wages | 2.4 | 5.1 | 2.6 | 5.2 |
| 3. | Expenditures for the maintenance of buildings and facilities, including current and capital repair | 6.4 | 13.5 | | |
| 4. | Transportation expenditures | 0.5 | 1.05 | | |
| 5. | Expenditures for fuel | 0.2 | 0.4 | | |
| 6. | Expenditures for the acquisition of inventory items of small value and short life | 4.8 | 10.1 | | |
| 7. | Quartermaster expenditures | 4.8 | 10.1 | 5.0 | 10.1 |
| 8. | Administrative: management and material expenditures for military education and scientific research conducted by institutes of the Ministry of Defense (MOD) | 1.9 | 4.0 | | |
| 9. | Expenditures for the creation, formation, and storage of reserves for military purposes | 2.1 | 4.4 | | |
| 10. | Expenditures for science | 2.5 | 5.3 | | |
| 11. | Expenditures for financing the expansion of subsidiary economic organizations of the MOD | 0.2 | 0.4 | | |
| 12. | Postal: telegraph and telephone expenditures | 0.5 | 1.05 | | |
| 13. | Expenditures for preinduction training | 0.1 | 0.2 | | |
| 14. | Expenditures for cultural and sports measures | 0.1 | 0.2 | | |

TABLE 4.2 CONTINUED

| No. | Title of articles | Previous reporting period, 1969 | | Reporting period, 1970 (expected fulfillment) | |
|---|---|---|---|---|---|
| | | Amount | Percentage of Total | Amount | Percentage of Total |
| 15. | Unidentified | 0.3 | 0.6 | | |
| 16. | Other expenditures | 0.7 | 1.5 | | |
| | Total | 47.5 | 100 | 49.6 | 100 |

[a] This is the section title. This table was included in a larger document, "Estimate of Expenditures for the Financing of Nonproductive Goals (Otsenka raskhodov na finansirovaniya neproizvodstvennykh tseley)."

As we discussed earlier, a small adjustment had been made in the 1955 prices to take account of the mid-1967 Soviet reform of industrial prices. Exactly how much prices for defense hardware were raised in mid-1967 is still unclear because the 1955 prices underlying CIA's estimates were themselves too low on average. Although the necessary correction was slow in coming, it was finally carried out.

The public reaction to the revision was also somewhat delayed, but when it came it gathered momentum. In May 1976, DCI George Bush appeared before Senator Proxmire's subcommittee of the JEC and discussed the new estimates. "Some of these changes," he told the subcommittee, "seem quite startling, certainly at first glance they do."[12] The declassification and publication of the testimony took months. But in early 1977, news of the revision found a wide audience. Two of the most prominent commentaries were a *Business Week* article ("The CIA's Goof in Assessing the Soviets")[13] and a Joseph Alsop piece in *The Washington Post* ("Cautionary Tale").[14]

The *Business Week* article quoted Graham, director of DIA at the time, to the effect that only DoD intervention prevented CIA from dismissing the emigré's information. It also gave an inaccurately high figure for the amount of the increase in the ruble value of procurement and said the sheer size of the adjustment "create[s] a strong presumption that the error was not limited to the CIA's underestimate of ruble prices in the Soviet defense sector. Quite possibly, more fundamental errors are involved, such as underestimating the quantity or performance capabilities, or both, of Soviet weapons systems" (although the emigré provided no information on either quantity or performance). Alsop also attacked the agency for attempting to squelch the emigré's report while crediting Graham with saving the situation. The agency's blunder in measuring Soviet military spending, he charged, was the result of CIA's hiring "broadly speaking . . . the American professorate," many of whom "have the ideological slants—often in extreme form—of any characteristically liberal American university professor."

That same year Eugene Rostow, in a foreword to one of William Lee's books published by the National Strategy Information Center, wrote "still, the tale of our intelligence estimates of Soviet defense programs in recent years will rank high in the literature of such pathological phenomena—close to the history of Galileo, surely, or the controversy over Darwin, or the theory of the four humors."[15] A few years later Richard Nixon complained that "American presidents were being supplied by CIA with figures on Russian defense spending that were only half of what the agency later decided they had been. Thanks in part to this intelligence blunder, we will find ourselves looking down the nuclear barrel in the 1980s."[16]

The consternation sparked by the revision reverberated elsewhere in the administration and in Congress. Appearing before an advisory panel of outside experts chaired by Ivan Selin—hereafter referred to as the Selin panel—David Chu, assistant secretary of defense for programs analysis and evaluation, spoke of the increasing distrust of the spending estimates, "particularly its relationship to Soviet GNP."[17] Also in testimony before the Selin panel, James Locher, a staff member of the Senate Armed Services Committee, highlighted the problem caused by major shifts in the estimates ("as in 1976") in his summary of criticisms of CIA's estimates by senators on his committee.[18]

In the academic world, most observers recognized that the revision should not affect judgments on Soviet military capabilities. Franklyn Holzman, one of the agency's most persistent critics, summarized the significance of the revision accurately—that it meant that the Soviets were devoting more resources to defense than had been believed earlier.[19] However, the credibility of CIA's estimates was questioned as it had not been before. Holzman asked why, given what CIA had said earlier about the margins of error in its estimates, outsiders should believe the new ruble figures or even the old dollar estimates.[20] University of Birmingham's Philip Hanson agreed that the estimates were shaky and hoped that until future estimates of the Soviet defense burden could be shown to be more accurate "policy implications (of these measures) will be ignored."[21]

Estimates

## Institutional Response

The 1976 revision of the ruble estimate and its aftermath triggered an institutional response that drove CIA's military-economic analysis for the next several years. The response encompassed an agenda for new work, charges to external review panels, organizational changes, and computational and methodological refinements and improvements.

### Research Agenda

First of all, although it was clearly a major improvement, the new ruble price base was anchored to only about thirty-five prices. These prices in turn were used to derive ruble-to-dollar ratios on the basis of estimated dollar costs that varied in quality. OSR pressed the intelligence community's collectors—especially NSA and CIA's directorate of operations—to uncover more ruble prices and product specifications. It also stepped up its program of contracts with U.S. defense firms for estimates of the dollar costs of Soviet weapons. During the next six years (FY 1976 to FY 1981), $2.1 million was budgeted by CIA for external costing contracts. Another $2.5 million was set aside in the following six years (FY 1982 to FY 1987). Contractors included leading firms and laboratories in the defense industry: RCA, TRW, Hughes Aircraft, Pratt & Whitney, Lawrence Livermore, Sandia, Raytheon, and many others. A 1988 primer on defense costing summarized the status of the external contract work (table 4.3).[22]

The external costing contracts were a joint CIA/DIA enterprise, and the CIA funding was for all practical purposes matched by DIA contributions. Each year a CIA/DIA Military Costing Review Board appraised the adequacy of the cost estimates of Soviet weapons systems in current use and recommended areas where external contracts could improve the estimates. In this arena, CIA/DIA cooperation proceeded smoothly.

Many of the cost factors for estimating operations and maintenance spending had their own weaknesses. As more intelligence on operating rates became available, estimates of spending for O&M—the costs of fuel usage, for example—could be refined. In what turned out to be a breakthrough, OSR began to exploit several Soviet field manuals that permitted the reestimation of a number of O&M line items.[23]

The least convincing part of CIA's overall estimate was spending for military RDT&E. Through the 1960s and early 1970s it was based on analysis that Nancy Nimitz, a senior researcher at RAND, had performed on Soviet-published material related to science. Her analysis depended on the ma-

TABLE 4.3
DATES OF LATEST DOLLAR COST STUDIES

| Weapons category | Date of latest cost study | Cost studies to be completed in 1988 |
|---|---|---|
| Land arms | | |
|   tanks | 1979 | X |
|   armored personnel carriers/ | | |
|     infantry fighting vehicles | 1981 | |
|   artillery | 1980 | |
|   multiple rocket launchers | 1985 | |
|   antiaircraft artillery | 1974 | |
|   trucks | 1976 | X |
| Naval | | |
|   ships | 1982 | X |
|   submarines | 1980 | X |
| Aircraft | | |
|   airframes | 1984 | X |
|   engines | 1987 | X |
|   avionics | 1986 | X |
| Strategic missiles | | |
|   airframes | 1976[a] | X |
|   engines | 1976[a] | X |
|   guidance | 1976[a] | X |
|   reentry vehicles | 1986 | |
|   ground support equipment | 1976[a] | X |
| Short-range ballistic missiles | 1983 | |
| Cruise missiles | 1986 | |
| Air-to-surface missiles/tactical air-to-surface missiles | 1986 | |
| Surface-to-air missiles | 1976 | X |
| Antitank guided missiles | 1985 | |
| Air-to-air missiles | 1976 | X |
| Space | 1984 | X |
| Electronics | 1978 | X |
| Nuclear weapons | 1984 | |
| Ammunition | 1978 | X |

[a]The 1976 study was reviewed and updated in the summer of 1986.

nipulation of open-source aggregative statistics and was therefore conceptually different from the building-block estimates of other defense accounts. Two attempts to put RDT&E on a building-block construct in the early 1960s and 1970s ground to a halt before completion. Attempts to gather enough information on individual Soviet institutes to permit building-block estimates proved unproductive. Within the mass of detail were nuggets of value for other analysis, but no end was in sight in terms of compiling an estimate of total spending for military RDT&E. When the agency revised its estimate of RDT&E in 1976, recognizing that total RDT&E in the Soviet Union was greater than official statistics on science expenditures, the new estimate was not considered much more reliable than the old.[24] The problem of military RDT&E was all the more serious because it was a large and growing share of total defense outlays. David Holloway noted that CIA estimates for 1967–77 put military RDT&E at 20–25 percent of total defense and equal to 40–50 percent of procurement.[25] These shares seemed high to him when compared with U.S. ratios of 10–12 and 25–30 percent respectively.

MEAP persistently urged OSR to upgrade its work on RDT&E and to warn consumers of the fragility of the existing estimate.[26] Through 1982, however, not much had been accomplished toward devising a more reliable estimate. At that point another attempt at a bottom-up estimate began. By the end of 1984 MEAP was reporting to the DCI that progress was being made on an improved methodology; in November 1985 MEAP reported that it was pleased to see how far the new methodology had come.[27]

To document the changes being introduced in the defense cost factors, a new methodology handbook was completed in 1976.[28] The previous handbook, published in 1960, was badly out of date. Numerous changes had been made in everything from division manning factors to costs applicable to the maintenance of barracks, but the documentation for these changes was dispersed in analyst files. The process of drafting the new handbook unearthed some anomalies and identified weaknesses that required more research to correct. In the late 1970s and early 1980s an even more extensive documentation was carried out. The result was an eight-volume compilation of methodological working papers covering CIA's estimates of Soviet military spending.[29] In these years each methodology statement was signed by the responsible analyst and a reviewing officer.

## Charges to Review Panels

MEAP played a prominent role in post-1976 reactions focused on the quality and value of Soviet defense spending estimates. Thus, in its report to DCI Stansfield Turner in August 1977, MEAP answered a question on the utility of the dollar comparisons asked by the director of OSR and the National Intelligence Officer for Economics.[30] In the same report MEAP took on broader issues related to CIA's work on the Soviet Union. It deplored the erosion in basic work on Soviet economic and political affairs (more information but fewer people working the problem) and told the DCI "a workable organizational solution to integrating specialized Soviet studies has not yet been achieved." Later MEAP suggested the formation of an "office of soviet studies" to effect the integration. The 1978 report of the spring MEAP meeting recognized a greater need for a multidisciplinary approach to Soviet military force issues "than we see positive steps being taken to accomplish it."[31]

During 1976–90 MEAP repeatedly pressed the importance of reworking the RDT&E estimate, doing a ruble comparison of defense spending to complement the dollar comparison of the size of U.S. and Soviet military programs, developing measures of the military capital stock, especially weapons inventories, and pursuing alternative methodologies to the estimation of aggregate Soviet military spending in rubles. When the issue of a slowdown in the growth of Soviet outlays was raised, MEAP urged that a retrospective on military developments in the 1970s be undertaken to try to understand the political and economic influences at work. It also supported the increased emphasis on research on Soviet defense industry.

After William Casey was appointed DCI in early 1981, MEAP provided him with an overview of the state of military-economic analysis on the Soviet Union.[32] Speaking to the dollar comparisons, MEAP said it was impressed with the robustness of the dollar estimates but that confusion persisted among users concerning their meaning. The agency, it suggested, should publish an unclassified paper explaining, for example, the conceptual differences between costing forces and programs and costing capabilities. While MEAP found no fault with the agency's concept of the ruble costs of Soviet defense spending, it pointed to "two serious problems in practice." First, the 1970 price base was becoming increasingly outdated in its representation of current trade-offs in the Soviet economy. Although the USSR intended to carry out a price reform in 1982, which theoretically at least would rectify the problem, it would be some time before the agency could acquire enough new prices for defense goods to convert its estimates to a 1982 price base. Second, the agency needed to take care in revising the ruble-to-dollar ratios that determined so much of the ruble estimate. The panel recommended that outside experts be convened to review the problem before the agency embarked on a "challenging and perhaps costly effort" to update the ruble-to-dollar ratios. In addition, it suggested that OSR warn consumers that a switch to a new price base could alter CIA's calculation of the Soviet defense burden.

DCI Casey and his new Deputy Director for Intelligence, Robert Gates, distrusted the agency's estimates of Soviet defense spending. In July 1982, DDI Gates wrote Selin, MEAP's chair, that it was time to undertake the broad review he had mentioned at the May 1982 MEAP meeting.[33] By November the Working Group on Soviet Military Economic Analysis (a subset of the full MEAP membership identified as the Selin panel) had formulated a work plan. Selin chaired the group, which separated into two subpanels—one to deal with the uses of the estimates (also chaired by Selin) and the other to review questions of methodology (chaired by Abraham Becker). The two panels took testimony from CIA analysts, policymakers at the Departments of State and Defense, congressional staffers, and outside experts (including almost all the major American critics of the estimates). The users panel met seven times from November 1982 to April 1983, and the Methodology Panel met thirteen times from November 1982 to March 1983. The working group's final report was released in July 1983.[34]

The working group was tasked by Gates to ponder three questions: How good are the current estimates of Soviet military expenditures and how can they be improved? How are the estimates used and how can they be made more useful? And, given the intrinsic uncertainties in the estimates and the uses to which they are put, would it be better not to publish some (or all) of the estimates?

In its final report the working group's principal findings regarding the way estimates were used can be summarized as follows:

- *A "truly amazing lack of understanding" existed regarding what the estimates represent, how they are developed, and how they should be used.*
- *The estimates have important uses not generally understood—they force analysts to confront questions that otherwise would be neglected such as "maintenance policies, ammunition stocks, production, and the mobilization base." The accounts also support interesting broad comparisons between U.S. and Soviet forces—for example, breakdowns by mission*

*and theater of operations. Nor could analysis of the Soviet economy be done without an appreciation of trends in defense outlays.*[35]

- *Many users exploit the estimates in a way that reflects badly on CIA's credibility, as in the political battles over U.S. defense spending. This finding, however, did not mean that the estimates should not be published ("an idea . . . both undesirable and impractical"). Rather CIA should "do a better job of explaining, documenting, and qualifying the estimates."*

The findings with respect to methodology covered thirteen points.[36] Beginning with an overall judgment that "CIA does an excellent job of estimating Soviet military expenditures," the panel endorsed the concepts underlying the dollar estimate, called the introduction of an alternative methodology for the RDT&E estimate "of the highest priority," rejected various criticisms of the estimates, and worried that estimates of spending in a current year were too provisional to be given the same credence as estimates for periods of three or more years past. The price base for the ruble estimates needed updating, which required a greater emphasis on gathering prices and understanding Soviet price formation. As MEAP had earlier, the Methodology Panel recommended that ruble as well as dollar comparisons be calculated and that research on alternative methodologies be given greater priority.

One finding touched on a question that was never resolved. Pointing to the procurement slowdown that was then a divisive issue in the intelligence community, the Methodology Panel argued that CIA's procurement costing methodology amounted to pricing the output of defense industry in fixed prices, with some allowance for learning curves. Therefore, the agency's estimates of procurement could only be interpreted as an index of real costs when inputs into defense production grew at the same rate as output.[37] To this day there is not sufficient information on the behavior of costs in Soviet defense industry over time to show how big a problem this might have been.

After examining the organization of military-economic work in the recently established Office of Soviet Analysis (SOVA), the Methodology Panel agreed that a major part of the rationale for the fundamental reorganization of the DI from functional to regional components—a change earlier endorsed by MEAP—was to encourage cross-cutting analysis. But it warned against an excessive diversion of analysts from what it called "component analysis," of which SOVA's military costing work was an example.[38]

SOVA, according to the panel, needed a single focus for its military-economic work in order to provide the "centralized methodology, discipline, and continuity that characterized these estimates in the past." Finally, the panel expressed support for SOVA's effort to update SCAM with a new, more flexible system that among other things allowed interactive data entry and editing.

The early 1980s witnessed a proliferation of committees convened to critique agency analysis of the Soviet Union. Gates commissioned Henry Rowen, former head of the National Intelligence Council (NIC), to set up an expert panel to "explore the hypothesis that the current state, or the future prospect of the Soviet economy is worse than estimated by the CIA."[39] In their deliberations, the Rowen group considered CIA's estimate of the share of defense in GNP and concluded that even the agency's broadest definition of Soviet defense-related programs was too narrow. It suggested a more expansive definition that encompassed "most of the Soviet Union's resource allocations for the conduct of external imperial activities." According to Rowen's working group report, a rough calculation of the costs of empire and other national security-related outlays would raise the Soviet defense burden to 20–23 percent of GNP.[40]

While the Selin working group was conducting its study, the President's Foreign Intelligence Advisory Board (PFIAB) considered the Soviet Union's deteriorating economic situation and its implications for U.S. foreign policy. It asked for an independent review of the intelligence community's analysis, and the NSC asked the Department of Commerce to undertake the task. Lionel Olmer, commerce undersecretary for international trade, supervised the study. Although the study's mandate specifically excluded an appraisal of CIA's estimates of Soviet military spending, it—like the Rowen panel—addressed the Soviet "burden of empire." It found no evidence that the agency's estimates of these burdens were "seriously in error."[41]

All in all, CIA's Soviet economic and defense-economic estimates received a continuous and careful scrutiny by outside observers after the mid-1970s. To our knowledge, no other aspects of agency analysis were accorded the same degree of independent expert review. As we explain later in this chapter, we believe this exposure did much to improve the estimates.

### Organizational Initiatives

Even before the 1976 upward revision in the ruble spending estimates, it was clear that a substantial increase in analytic resources was required to repair the long neglect of the ruble price side of the costing database. The revision created the bureaucratic shock necessary to bring about the increase. In April 1976 OSR's Programs Analysis Division, the component responsible for producing the spending estimates, announced an "important milestone"—the beginning of planning for twenty new positions in the office's military-economic effort.[42] By the end of 1979, Rae Huffstutler, OSR's director (and later the first director of SOVA), could report that he had thirty-seven analysts working on Soviet military-economics in 1979 and 1980.[43] The Programs Analysis Division, which had existed before OSR was established, had three branches. In late 1977 the Military-Economic Analysis Center was formed with four branches (Defense Industries, Economic Implications, Comparative Analysis, and Manpower and Operations) and an analytical support group to manage the SCAM costing model.

When the DI later reorganized from a functional to a geographic orientation in late 1981, military-economic analysis was concentrated in two divisions. A Defense Industries Division had three branches responsible for industrial analysis, defense production estimates, RDT&E, and management of external costing contracts. The Econometric Division's charter covered military-economic and some economic accounts. It had branches for comparative analysis (dollar costs of Soviet programs), defense and economic accounts (ruble estimates), and growth and forecasting (measures and econometric modeling of Soviet economic performance). The division provided measures of growth of Soviet GNP and defense, as well as comparisons of U.S. and Soviet GNP and defense.

Assessments of the new organization in SOVA differed. Burton told the Selin panel that defense costing was an extremely complex undertaking requiring constant, careful oversight to avoid another 1976 incident. In terms of capability, he rated the new organization as "no more than 20 percent as good as the old organization."[44] From another perspective, Noel Firth, the former deputy director of OSR, cited problems resulting from the division of responsibility for Soviet economic matters between OSR and OER in the old functional DI organization. He thought military-economic analysis should work better in the new SOVA format.[45]

In 1985 the two activities—following defense production and analysis of defense-economic policy with the help of the defense-costing work—were brought under one tent with the creation of the Defense and Economic Issues Group in SOVA. Two divisions were subordinated to the group—the Defense Economic Division and Defense Industries Division. The Defense Economic Division, as before, dealt with costing, defense

Estimates

policy, economic and defense measures, and economic forecasting. Its three branches were also the same: defense comparisons, economic implications, and defense economic policy. The Defense Industries Division had branches for defense management, production analysis, dollar costing of procurement items, and industrial technology. The Defense and Economic Issues Group had about sixty-five people during the late 1980s.

Although the names of the branches didn't change much, the focus of the work did with Mikhail Gorbachev's selection as politburo chair. The Defense Economics Division grappled with the question of how Soviet economic problems and Gorbachev's economic plans would influence defense programs. The Defense Industries Division's study of high-technology sectors in the Soviet Union was accorded greater prominence (and more analysts), in part due to Casey's interest in the subject. The Defense Management Branch concentrated on the implications of Gorbachev's industrial modernization drive for defense industry and defense production and presided over the introduction of a new methodology for estimating Soviet RDT&E and its military component.

### Computational Improvements

In 1971 the computation of the ruble and dollar costs of Soviet defense programs was redesigned to run on the IBM 360 computer. By then, however, it was evident that SCAM's costing techniques and data organization could not handle some of the questions addressed to OSR. In a joint effort, OSR and CIA's Office of Data Processing began to develop a new system called SCAM II. Developmental work was finished in March 1974, but for a variety of reasons the new model was not ready for complete testing until April 1976. At that time a lengthy debugging process, involving successive comparisons of results from applying SCAM and SCAM II to the same input data, was begun. This process uncovered many problems that were ultimately resolved, and the new system was first used operationally for the 1977 update of the spending estimates.

The basic features that distinguished SCAM II from SCAM I were an option for provision for order-of-battle generated production estimates; ability to break out expenditures, manpower, and force levels by service, commands, and region; an expanded machine-readable file of the output data (expenditures, order-of-battle, production, and manpower); an improved input capability allowing analysts to enter data more easily; a price index program that facilitated base-year price changes; capability to produce expenditure reports in current and constant prices; a unit detail summary report that brought together all cost factors and production and order-of-battle data in a single report; an activity factor that allowed O&M to be generated as a function of unit detail activity; and an option to compute O&M as a function of military investment.

### Major Changes in Evidence and Methods

The last fifteen years of the defense spending estimates were marked by important additions to the stock of knowledge supporting the estimates and advances in exploiting the available information—often through the application of statistical techniques. Most of the changes were the result of more people working on the problem following the 1976 estimate revision. Other changes reflected improvements in collection methods and increased attention to defense-economic requirements in tasking collection systems. The 1961–75 neglect of the database had ended. From 1975 to 1990, new information and analyses were used to improve the estimates for major resource categories of spending—personnel, O&M, procurement, construction, and RDT&E. We discuss the most important of these improvements in detail in appendix A.

## Two More Controversial Changes in the Spending Estimates

Between the 1976 jump in the ruble estimate and the 1991 breakup of the Soviet Union, two additional controversial changes were made in CIA's estimates of Soviet defense spending. The first—a delayed call on a plateau in Soviet spending for procurement—caused consternation in Washington's policy community; we discuss it in the next section. The second—the conversion of the ruble estimate from 1970 to 1982 prices—raised the share of defense in GNP and aroused skepticism in the academic community. We discuss this change in detail in chapter 6, which focuses on academic and other external criticisms of CIA's military-economic intelligence.

### Slowdown in the Growth of Defense Spending

In February 1983 CIA published a dollar-comparison paper explaining that "this year's review shows a period of almost no growth in the dollar costs for Soviet procurement from 1976 to 1981." The agency attributed the estimate change (see table 4.4 for a history of previous estimates) to changes

TABLE 4.4
RECORD OF PUBLISHED ESTIMATES
OF SOVIET DEFENSE SPENDING, 1977–83[1]

| Year estimate completed | Year estimate published | Last historical year for data[2] | Reported trend for late 1970s and early 1980s (for early papers this is a forecast; for later papers it is a combination of historical data and forecast) |
|---|---|---|---|
| Defense spending (in ruble terms) | | | |
| 1977 | 1977 | 1976 | Total spending grows at 4–5 percent per year |
| 1978 | 1978 | 1977 | Total spending grows at 4–5 percent per year |
| 1979 | 1979 | 1978 | Total spending grows at 4–5 percent per year |
| 1981 | 1981 | 1980 | "Growth in procurement slowed in the late 1970s . . . this reflected cycles in procurement . . . rather than signaling a new trend." |
| 1982 | 1983 | 1981 | No growth in procurement since 1976; new trend acknowledged |
| Cost comparisons (in dollar terms) | | | |
| 1977 | 1977 | 1976 | — — — |
| 1978 | 1978 | 1977 | — — — |
| 1979 | 1979 | 1978 | Investment leveled off in 1977 but grew again in 1978 |
| 1980 | 1980 | 1979 | One year of no growth (1977) in procurement but turns up again due to procurement cycle (strategic weapons) |
| 1981 | 1981 | 1980 | Two years of no growth (1977 and 1979) in procurement, but it turns up again due to procurement cycle (aircraft) |
| 1982 | 1983 | 1981 | Four years of no growth (1977–80) in procurement |

[1]Extracted from CIA/SOVA, "Record of Published Estimates of Soviet Defense Spending," Memorandum for Deputy Director of Intelligence, January 25, 1985.
[2]The end year of the historical period of the estimate may contain false growth because the methodology takes into account the lead costs for specific systems that are expected to be (but may not be) deployed in future

TABLE 4.4 NOTES CONTINUED

years. It must be considered less well supported by underlying evidence than in earlier years. Therefore, the end year must be interpreted with caution.

"in our assessments of the pace and timing of selected weapons in the classes that make up the bulk of procurement—specifically, ships and boats, aircraft, and strategic missiles." New evidence that "became available in 1982" led to the adjusted assessments. The result was a "procurement plateau" that had lasted longer than previous cyclical change in procurement and that might "signal a secular change in the pace or composition of military procurement."[46] The new estimate showed substantially slower growth than the previous estimate in total defense spending after 1976, in large part because of the plateau.

The news of a slowdown had a sour reception in policymaking circles. In November 1982, a SOVA analyst recalls, CIA was asked to prepare some material on Soviet military spending for a speech by President Ronald Reagan. The statistics were intended to support continued growth in the U.S. defense budget. By this time SOVA analysts had concluded that the SCAM estimates did not jibe with the steady 4–5 percent growth in Soviet military outlays described in earlier CIA papers, and they so informed SOVA Director Huffstutler. An embargo was placed on the spending numbers, as Huffstutler ordered an officewide check and double check of the estimates to ensure that the perceived slowdown was adequately supported by physical evidence.[47] After the slowdown was confirmed, SOVA's Econometrics Division submitted a traditional ruble spending paper to Huffstutler for publication approval. Meanwhile, the Policy Analysis Division had prepared a draft paper speculating on the implications of the slowdown. Huffstutler ordered the two divisions to merge the drafts in a paper that appeared in July 1983.[48]

Before it was published, the paper was briefed in the NIC and to Secretary of Defense Caspar Weinberger. DCI Casey brought the news to President Reagan, in effect saying "we have a problem" at a time when Weinberger was testifying before Congress about the need for a higher defense budget.[49] The draft encountered strong resistance from the National Intelligence Officer (NIO) for General Purpose Forces, Maj. Gen. Edward B. Atkeson, and his assistant.[50] George Kolt, Assistant NIO for the USSR (later director of SOVA and thereby responsible for producing the

Estimates

spending estimates), was also highly critical.[51] The NIO-at-large suggested that growth in defense spending in terms of real resources had not really slowed. Rather, bottlenecks in production and other factors had caused cost overruns.[52] The secretary of defense "went nuts."[53] One of Weinberger's deputies, Fred Ikle, told the agency of his particular concern that the CIA's cost estimates showed declining outlays for Soviet strategic intercontinental forces (40 percent from 1974 to 1981).[54] So the controversy continued through most of 1984, although the focus shifted to the reasons for and likely duration of the slowdown rather than its existence—at least through 1981.[55] Gates expressed his frustration with the whole affair in a memorandum to the DCI and the deputy director of Central Intelligence (DDCI): "I think all three of us probably wish this whole costing effort would just disappear because I don't know of a single substantive subject that breeds as much heartburn and as little positive return."[56]

CIA's July 1983 paper, *Soviet Defense Spending: Recent Trends and Future Prospects,* did not identify a single or even a primary cause of the slowdown. Instead it pointed to a constellation of possible explanations:[57] technical delays in assimilating new weapons into production, bottlenecks in the supply of raw materials and energy and transportation tie-ups, and a policy decision to curtail defense spending, particularly on procurement. There was some evidence of technical delays, and bottlenecks and transportation difficulties had been widely noted in open and classified sources. The possibility of a policy decision was advanced with considerable timidity: admitting there was no direct evidence that Moscow "either anticipating these problems, or in response to them, may have taken steps to stretch out some procurement programs."

Two SOVA analysts were directed to further explore the reasons for the slowdown. In early 1985 they delivered a draft paper concluding that the plausible interpretation was that the USSR had planned a more selective (constrained) force development in the 1976–80 five-year plan and in the initial years of the 1981–85 plan. The draft paper argued that by the mid-1970s the short-term military threat from the West seemed manageable to the Soviets with the attainment of strategic parity. The long-term threat posed by western advanced systems, the paper suggested, inclined the Soviet leadership to pay more attention to investing in the defense-industrial infrastructure and Soviet technological capabilities. At the same time, the long slide in overall Soviet economic growth provided a powerful reason to moderate the growth in defense programs.[58] This draft, however, and subsequent revisions never satisfied SOVA managers who reviewed it. The analysts suspected the message was politically unacceptable during a period when the administration was trying to sustain the momentum of the U.S. defense buildup, while the managers thought the case for a deliberate decision on the part of the USSR to curtail the growth in defense spending simply had not been made convincingly enough. Part of the problem may have been that managers had a different standard of proof for a slowdown than they did for continual growth in procurement. Even though the expected cyclical upturn in procurement outlays failed to materialize year after year, these expectations were given more weight than the recent historical record.

Meanwhile critics of administration policy seized on the plateau as evidence that U.S. policy was wrong. Raymond Garthoff wrote that talk of a "relentless Soviet buildup" was mistaken and that the "spending gap" was no more solid than other gaps (bombers, missiles, antiballistic missiles, civil defense) that had been headline news.[59] In a later analysis of Soviet foreign policy, Garthoff emphasized that the Reagan administration's spending gap was "deflated by its own intelligence estimates."[60] Senator Orrin Hatch, in a letter signed by six other senators, wrote Casey about press reports that NATO had endorsed CIA's estimate regarding slower growth in Soviet defense spending. He said, "We are sure you appreciate the difficulty in persuading our colleagues in the Senate to support the president's request for a 10 percent real growth in our defense budget a few weeks after NATO and the U.S. Intelligence Community have apparently endorsed a 50 percent 'cut' in estimated Soviet defense spending."[61]

After questions arose concerning the reasons for the delayed call on the slowdown, the CIA explained that under normal conditions (i.e., good data and conventional methods), three or four years of flat procurement might give "a decent basis for claiming that a change in trend had actually occurred."[62] It added that it had hesitated to announce a slowdown in the growth of total defense outlays or a procurement plateau because the military analysts who had lived with the Soviet buildup of the 1960s and early 1970s were conditioned "to expect more of the same." Furthermore, the large number of defense programs in research and development (R&D) and the expansion of floor space in defense production facilities led analysts to believe procurement would continue to grow.[63] On the other hand, the spending estimates in the late 1970s did not confirm the continued growth in procurement and aggregate defense programs that analysts anticipated. In a major paper on Soviet defense programs published at the beginning of the Reagan administration, OSR waffled on the issue. It described trends in total defense spending in general terms without mentioning procurement. In the Brezhnev years, the paper said, defense spending had increased at an average of 4–5 percent per year, "with the growth slightly more rapid in the late 1960s than in the 1970s." CIA's baseline projection for the 1980s was lower—2 to 4 percent per year—largely because of inferred Soviet concern over the long-term costs to the economy of continued growth in defense outlays at previous rates.[64] (In chapter 6 we consider whether Soviet leaders would or could have seen a reduction in the rate of growth of inflation-adjusted defense spending.)

As participants reconstruct the issue of the delay in the announcement of the slowdown, one factor stands out. In the expenditure estimates for a given year and the preceding year, outlays for the lead costs of major procurement items—for example, ships, missiles, and aircraft—whose production is *anticipated but not yet observed* are important. With each year's new estimate, procurement in the preceding and current years tended to be higher than in the previous years. Some time in 1981 or early 1982 analysts recognized that these repeated upturns in procurement depended heavily on the lead costs of future and as yet unseen systems and that, unlike the past, these "future systems had not been materializing at the times and in the magnitudes predicted."[65] At a MEAP meeting in the fall of 1982, one MEAP member looked at the latest spending estimates that were being projected on a screen and pointed out that they were not consistent with SOVA's reporting of a persistent 5 percent growth in real Soviet defense procurement. Huffstutler agreed and set in motion the detailed reexamination.[66]

### *Conversion of the Ruble Estimates to a 1982 Price Base*

In 1982 the USSR conducted a major price reform that made it logical to convert the defense spending estimates from a constant 1970 ruble price base—the period reflecting the last Soviet major overhaul of prices—to a constant 1982 ruble price base. Easier said than done. Over the next few years the intelligence community collected a sample of ruble prices of military hardware—which were extremely difficult to obtain—and CIA developed some ingenious techniques to fully exploit this newly acquired price information. By 1986 it made the conversion to 1982 prices and derived the following principal conclusions: substantial inflation had occurred in the prices of military hardware between the last major price reform (1967–70) and the 1982 price reform; inflation in the defense sector was greater than in the economy generally, so that despite slower growth in defense, the defense share of GNP remained at about 14 percent whether measured in 1970 or 1982 prices; the plateau in procurement spending revealed in the 1970 ruble-based estimates also appeared in procurement estimates using 1982 ruble prices. The CIA conversion to 1982 prices was generally well received by users of the estimates, but it received strong criticism from outside researchers in the academic community. We discuss these criticisms and the CIA response, which details the methodology for making the conversion, in chapter 6.

## Moving into New Areas

During the 1980s CIA pushed its costing effort in four new directions. It developed spending estimates for an expanded definition of Soviet national security expenditures; it produced estimates of Soviet defense outlays expressed in current rubles; it worked out a new approach to project likely future military forces; and it moved beyond the U.S.-USSR defense comparisons to present NATO-Warsaw Pact (WP) comparisons. The reasons for the controversy surrounding the estimates involving an expanded definition ("cost of empire") and the NATO-WP comparisons will be discussed in the next chapter. The current ruble estimates and the attempt to improve force projections are more conveniently treated here.

### *Estimates of Soviet Defense Outlays in Current Rubles*

Before the late 1980s, CIA's estimates of the cost of Soviet military programs had almost always been presented in constant rubles or constant dollars (prices of some base year). The objective was to measure real growth in these programs. But for the analysis of defense burden, a set of GNP accounts in current rubles, including defense, is the best starting point because the price changes reflected in current ruble accounts presumably reflect changes in the rate at which goods can be substituted for one another in production. In the Soviet case, where prices were not market determined and price reforms were infrequent, the reflection was always imperfect but probably better than the picture given by GNP accounts in constant prices.

Converting constant ruble estimates into current ruble estimates required estimates of price change for goods and services purchased by the Soviet military. The big stumbling block had always been the lack of an adequate price sample for defense procurement. With the expansion of the price database, an estimate expressed in current rubles was now feasible. The approach to measuring price change or inflation was essentially the same as that used in converting from a 1970 to 1982 price base—reliance on open sources where possible, classified sources to fill gaps, and regression analysis performed on the sample of procurement prices.[67] The results of the current ruble estimate are depicted in figure 4.2 for the period 1970–90. First compiled in 1988, the current ruble exercise was rerun in March 1991 when a new batch of prices became available.

## The Force Projection Problem

Historically CIA's projections of future military spending (and underlying forces and programs) tended to go flat or even decline compared with the estimates through the current year. This phenomenon was so common it was given a name—the "tired-arm" effect.[68] It was the result of analysts' failure to predict the appearance of new models of weapons or the continued production of existing models to the degree that later evidence indicated. Therefore, when projections in the late 1970s showed a decline in defense procurement, OSR management was skeptical, as the previous thirty years of experience indicated they should be.

Figure 4.2
Soviet Defense Outlays: Billion 1982 Rubles
and Billion Current Rubles

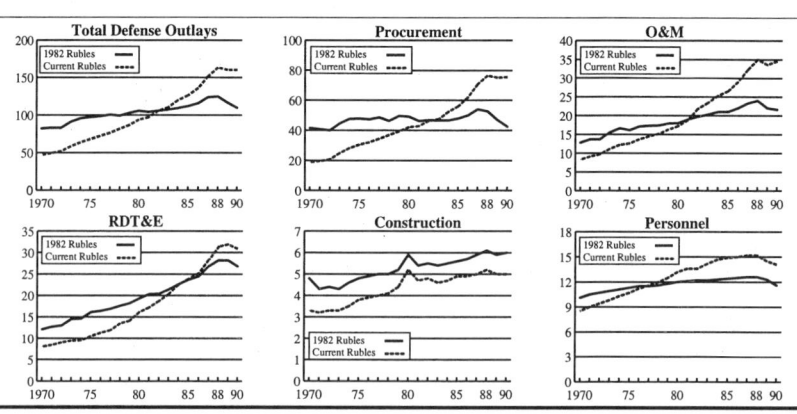

The reaction to a May 1977 internal memorandum illustrates the problem. The memorandum pointed out that the existing projections of procurement implied an absolute decline in spending for defense production in 1976–79, with a reduction in spending on ships and aircraft accounting for most of the fall. While recognizing that "we have historically underestimated the size of future forces," the author asked whether there might have been a major shift in Soviet defense policy.[69] The OSR branch chief immediately responsible for the projections of defense spending in rubles at the time recalls that he didn't believe the projections "because of (a) my experience and (b) probably more importantly the conventional wisdom around the unit that Soviet procurement is cyclical and we knew we had just finished a major wave of missile procurement."[70] A related briefing paper on the 1977 projections concluded that past projections of Soviet procurement spending had been "relatively useless," in large part because of inaccurate projections of aircraft and land-based missile production.[71]

To improve projections of defense production, the briefing paper concluded that OSR must better take into account available production capacity and demonstrated Soviet willingness to procure large numbers of new and upgraded weapons, in addition to order-of-battle data.[72] The briefing announced that the projections of physical production OSR's military analysts would make in 1978 would be reviewed by the military-economic analysts for consistency with past trends in total spending, allocations by military service, and available defense production capacity. Meanwhile OSR adopted a temporary fix by applying a 3 percent additional growth factor to all projections of spending for defense procurement.[73] This arbitrary correction factor eliminated the tired-arm effect, but the spending projections it produced turned out to be wrong in the opposite direction. They repeatedly predicted upswings in procurement that did not occur. Moreover, because of the lead costs of future weapons systems—that is, the costs counted in years before the year of actual deployment—adoption of this correction factor artificially raised estimated procurement in the years immediately preceding the projection period, thereby contributing to the delay in recognizing the post-1975 plateau in Soviet military procurement spending.

With increased attention directed to specific Soviet military RDT&E programs, the situation changed in the 1980s. The military analysts responsible for making projections of production now had an abundance of evidence of developmental programs intended to result in series military production. They were encouraged to think creatively about future production of weapons and equipment, with this RDT&E evidence in mind, in order to offset the tired-arm phenomenon of the past. During the early 1980s, this procedure resulted, year after year, in an abrupt upturn in projected total procurement spending in contrast to the spending levels for the years immediately preceding the projection period. Instead of a tired-arm effect, the procurement spending projections now displayed a "ramp" effect. The system had overcorrected.

In 1985 CIA began to develop a new approach to the projections problem.[74] It started from the premise that a major part of the overstatement of procurement embodied in the projections was caused by the failure to take into account the near certainty that some weapons development programs would be delayed or canceled. All possible future production programs, in effect, were given an equal but unspecified probability of occurring. Moreover, the system did not explicitly differentiate the uncertainty of future projections from known historical information or between different projected programs. As an initial experiment, more than three hundred major Soviet military production programs being projected at the time were reviewed. For each program, the responsible military analyst provided confidence levels expressed as probability statements with respect to the level and timing of future production.[75] The production projections were used to generate a cost line, and computer simulations—reflecting the uncertainties specified for each cost line—were used to generate a distribution of possible outcomes for each weapon system. As it turned out, ten thousand trial projections gave a stable median or best estimate with a 90 percent confidence level.

This new projections methodology served as the basis for SOVA's projections of Soviet procurement spending from late 1985 on, but it was not used to support the NIE process. Compared with the previous projections methodology—the sum of individual analyst estimates without any explicit specifications of confidence levels—the probability-based methodology predicted much lower growth in total procurement spending (see figure 4.3). SOVA believed "the new method . . . provides a much improved assessment of future programs"[76] while recognizing that it did not deal with all sources of uncertainty in projecting Soviet defense production. In particular, SOVA acknowledged that major shifts in the political and economic environment could lead to changes in Soviet defense policy. When the projections paper was drafted, Gorbachev was launching his perestroika, and the paper declared that it was crucial to examine the resource implications of the force projections jointly with projections of Soviet economic

**Figure 4.3**
**Soviet Defense Procurement** [a]

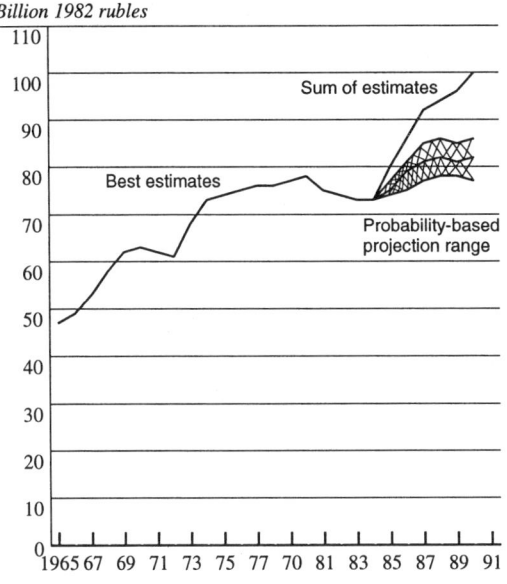

[a] These estimates were compiled in 1986 and differ from the more recent estimates shown elsewhere.

performance "to determine if adjustments are required in either or both estimates."[77]

The new methodology had one major drawback that probably would have prevented its use in the NIE process even if the Soviet Union had not broken up. It could forecast certain aggregate levels of procurement spending, such as spending for total procurement by branch of military service, but not item-by-item, year-by-year projections for individual Soviet weapons systems; this was often the basic information U.S. force planners wanted.[78] The NIEs, therefore, went on using estimates of projected deployment (and production) of individual weapons systems unconstrained by military-economic judgments. The sum of estimated expenditures for these estimates implied a much higher growth of defense spending than SOVA was predicting.

This disparity was strikingly illustrated in the preparation of NIE 11-3/

Estimates

8-86, the Soviet strategic forces NIE. In 1985, SOVA pointed out to the NIO for Strategic Programs (NIO/SP), who was in charge of drafting the NIE, that the cost implication of the force projections in the draft NIE were not credible. This view was ignored. SOVA then prepared a formal memorandum for distribution in the national security community reporting its findings. It explained that if the Soviet programs "proceed as currently projected," they would push the growth in annual outlays for strategic programs to 5–7 percent through 1989 and raise them to "the highest absolute level we have seen in two decades."[79] At the same time projected real spending on general purpose forces was slated to increase at 3–4 percent per year. These projections, the memorandum pointed out, would conflict with Gorbachev's economic modernization program. Moreover, technological and manufacturing problems made it unlikely that all of the programs would be as successful as the sum of the individual estimates implied.[80] The memorandum's impact on the debate was "attenuated," however, by the fact that when it was finally released its distribution outside CIA was limited to a few copies.[81]

In a last-ditch attempt to modify the language in the NIE, SOVA's director sent the DDI a memorandum proposing a dissenting CIA footnote to the NIE in order to force the intelligence community "to deal with the issue in the main text next go-around." The issue was SOVA's problem with the NIE force projections, which by then implied spending growth of 11 percent per year in strategic programs—even for "low forces." Such high growth, the memorandum argued, "would—unless accompanied by significant cuts in conventional military forces—not only rule out any prospect of success in the industrial modernization program, but in fact would imply that Moscow has no intention of attempting to carry out the program," the centerpiece of its economic agenda.[82]

Gates rejected the idea of a CIA footnote to the NIE. Instead the NIE declared the reverse—that economic considerations alone would not "lead the Soviets to abandon major strategic weapon programs, to forsake force modernization goals, or to make substantial concessions in arms control."[83] Indeed, they could and would increase spending substantially above the projected levels if they felt it necessary. The only concessions to economic constraints were a recognition that the rate of modernization of strategic forces "will be influenced by economic factors more now than in the past" and a footnote explaining that not all of the projected forces would appear on schedule "because of unforeseen problems in R&D or production." At the same time, the NIE added, some systems might be deployed faster than projected, and others, "not detected in development, will be deployed."[84]

The notion of either macro- or microlevel economic constraints on Soviet strategic programs was treated in the same throwaway fashion in NIE 11-3/8-87 (July 10, 1987). The summary explained that economic factors "might affect somewhat" the deployment of some strategic systems, but "we judge that strategic forces will continue to command the highest resource priorities" and therefore would be affected less by economic problems than other demands of the military. The NIE raised the possibility that enhanced competition for resources between Gorbachev's industrial modernization program and military requirements "could somewhat affect" the pace of some programs but said that the fact that the defense production facilities were already in place suggested minimal impact. As for technological and manufacturing constraints on the sum of the best estimates, the 1986 footnote was repeated: some programs might not appear as estimated but the misjudgments were likely to be offsetting. In short, the situation of the 1950s—when military-economic analyses had a significant impact on national estimative judgments—had been almost completely reversed.

## Managerial Support for Spending Estimates Erodes

### *Wider Use of Estimates Generates Distrust*

Through the early 1980s the demand for CIA's estimates of Soviet military spending continued at a high level. The dollar comparisons of the cost of Soviet and U.S. programs were especially in demand; they proved useful in supporting the U.S. defense buildup during the Carter and early Reagan presidencies. William Kaufman, a longtime Pentagon consultant, recalled that during the Nixon and Ford administrations he "turned very strongly" to the CIA's estimates "as a basis for indicating why the U.S. defense budget ought to be increased." To this end the comparisons were inserted into the annual DoD posture statements.[85] They also appeared in *Soviet Military Power*, the DoD series documenting various facets of the Soviet military threat.[86]

Indeed, the increasing use of the CIA estimates in the debate over the U.S. defense budget aroused suspicion about their integrity in Congress and academia. James Locher, a staff member of the Senate Armed Services Committee, told the Selin panel "that there are a number of people who believe this analysis is intentionally misleading, that it somehow comes out to exactly what it is that the Administration would like to argue on Capitol Hill. I mean they totally reject it." Depending on what the United States is doing with its defense budget, either side of the "political spectrum—(will be)—unhappy with the analysis."[87] Writing for the Stockholm International Peace Research Institute (SIPRI), Carl Jacobsen charged that the level of CIA's dollar estimates of Soviet military program costs coupled with high rates of growth of these costs "were seen to legitimize and compel the Reagan Administration's historically unprecedented defense budget increases."[88] Holzman asserted that "over the past 15 years, the CIA's estimates have been riddled with errors and misrepresentations, all making Soviet defense expenditures appear larger than they actually are."[89] On the other hand, Rosefielde wrote in 1982, "my impression is that the political pressures on the CIA to keep their estimates low are very strong."[90]

## *Agency's New Senior Managers Skeptical*

When Casey became DCI in 1981 he was not on record regarding CIA's estimates of Soviet defense spending. Early on, however, he indicated his doubts about the dollar estimates by asking Huffstutler how much could be saved by eliminating them.[91] Very likely, Casey's views were influenced greatly by those of Gates, who served him first as his special assistant, later as his DDI, and finally as DDCI. In 1984 when Gates was preparing to brief Weinberger on the latest dollar comparisons, which showed that the difference between U.S. and Soviet defense spending had narrowed considerably, he told Casey he planned to urge DoD to rely on physical comparisons rather than dollar comparisons of the defense programs.[92] According to meeting records, "the DCI agreed that this would be a better approach than using a lot of artificial cost figures."[93] Nonetheless, Casey kept his distance from decisions over the spending estimates and CIA's economic papers in general. Gates insists, for example, that Casey never interfered with the publication of any economic paper.[94] This claim is disputed by John Reynolds, a former manager of CIA's military-economic analysis, who reviewed this book in draft and noted that he personally drafted memorandums for Casey's signature refusing requests from the Senate Select Committee on Intelligence (SSCI) for public release of CIA's dollar cost comparisons, even though they had been made available in earlier years. Both Gates and Reynolds could be right in the sense that Casey at the outset clearly intended to stop the dissemination of unclassified CIA reports.

Gates proved to be the driving force in turning agency management against the defense spending estimates. When he became DDI in 1981 he already had "some fundamental misgivings about some of the CIA's work on costing of the Soviet defense effort" based on the fundamental difference in the U.S. and Soviet economies and his feeling that sufficient data did not exist to make accurate comparisons.[95] He distrusted the dollar comparisons as only an "analytical construct" and pressed to substitute physical comparisons (numbers of missiles, aircraft, and tanks) for the dollar comparisons. He also thought the ruble defense spending estimates did not take into account the many ways the Soviet economy was skewed toward military purposes. "No one around town," he said, believes CIA and DIA when they say defense is only 13–16 percent of GNP.[96] Instead Gates said that, when asked, he would answer that the Soviet economy was about 15–20 percent purely military and 20 percent purely civilian, and the remainder could not be broken down into civilian or military components.[97] Gates's intuitive views were generally shared by those who followed him as DDI. In fact, his appointment as DDI signaled a marked change in the background and attitude of the people who occupied that position. Whereas DDIs of the 1970s like Proctor and Clarke came from intelligence backgrounds of military and military-economic analysis, understood the estimates thoroughly, and felt comfortable explaining what they meant and didn't mean, DDIs of the 1980s didn't understand the estimates and were not at ease discussing or defending them.[98]

Because of his discomfort with the estimates, Gates decided to commission Selin, then chair of MEAP, to establish a working group of outside experts, including some MEAP members, to review the utility and reliability of the estimates. As we have discussed, the resulting working group made several recommendations for improvements in the estimates, but it fundamentally endorsed them.[99] The group's findings were reported to Weinberger and, according to Casey, received wide distribution within the administration.[100] Still, the intuitive discomfort with the spending estimates and the apparent lack of confidence in subordinate experts persisted.

## Estimates No Longer Helpful

The distrust of the estimates stemmed in large part from the lingering embarrassment over the 1976 major revision of the ruble estimate and the post-1982 controversy surrounding the announcement of the slowdown in the real growth of Soviet defense outlays after 1975. Gates delayed several months before approving the SOVA paper dealing with the slowdown, "awaiting the outcome of the Selin panel report and SOVA's work with DIA" (comparing CIA and DIA estimates).[101] In his forwarding memorandum to DCI Casey and DDCI McMahon, Gates said, "this is the paper over which I have had the greatest doubts," but he now found it persuasive because of SOVA's work on the Soviet economy and the endorsement of SOVA's economic and military-economic analysis by the Selin panel and the PFIAB panel, which he called "the two most far-reaching, comprehensive external evaluations of our analysis in many years."[102]

When the paper's findings became known, there was general consternation. The Senate Armed Services Committee, for example, canceled a major hearing because the idea of slower growth in Soviet defense spending generated enough controversy to make "our hearings less useful for our purposes."[103] Just as the concept of a plateau was being digested, DIA poured fuel on the fire by reporting in mid-1984 that Soviet procurement (valued in constant dollars) had increased by 5–10 percent in 1983 while CIA's preliminary estimate was 2–3 percent.[104] In testimony before the JEC, DIA was questioned intensely about this difference. CIA's representatives were grilled on the same discrepancy and testified that they disagreed with DIA on the estimates of physical production in 1983. They said the DIA estimate of spending growth was based on a smaller number of Soviet procurement items than CIA's estimates and that the DIA methodology overweighted higher growth items.[105] After a comparison of their respective estimates, DIA scaled back its estimate, but the 1983 upturn in procurement in the CIA and DIA estimates turned out to be caused by lead costs of future systems that were never procured at anticipated levels.

Meanwhile, the CIA/DIA differences were awkward for the intelligence community. In part they led Gates to ask whether the paper announcing the slowdown in Soviet defense spending should be delayed.[106] When the agency's classified dollar-comparison paper was published in March, the findings were leaked to the press—findings that were sharply at variance with DIA's estimates. To defuse headlines heralding a CIA/DIA split, Ikle arranged a March 1983 press conference involving DoD, DIA, and CIA to set the record straight. DIA and CIA reported their positions while minimizing their differences and cautioned that even with unchanged levels of spending, the USSR would buy enormous quantities of military equipment.

In 1986 when CIA and DIA testified jointly before the JEC, the two agencies still disagreed on trends in 1983–84 procurement.[107] Even so, the degree of convergence in views and the fact that the testimony was joint led some to charge that CIA and DIA had fudged their differences. In this vein, Holzman argued that CIA "may have moved, however reluctantly, to harmonize its published estimates of ME/GNP with those of the DIA."[108] Nothing so sinister occurred. The agency's estimate as presented to the JEC continued to show slow growth in real Soviet defense outlays until an acceleration forecast for 1985–88. The estimate of the burden increased because of the changeover to 1982 prices. Beginning in 1985, however, CIA/DIA consultations over defense-economic matters were held on a regular basis with a revival of the Joint Military Costing Review Board, which met quarterly to review estimates and other work in progress.

From some consumers' standpoint, the finding that the gap, measured in dollars, between higher Soviet and lower U.S. defense spending was narrowing also meant that the CIA estimates were becoming a liability. Gates, already dubious about the utility of the dollar estimates, briefed Weinberger in July 1984. In a draft memorandum summarizing the substance of the session with Weinberger, Gates referred to "future use of expenditure analysis by the Department of Defense" and noted "the use of the dollar comparison figures of Soviet defense expenditures by DoD to defend the defense budget was becoming less useful." He urged DoD to use physical comparisons of forces and procurement and noted that as the U.S. buildup continued, the U.S.-Soviet gap in spending would continue to narrow "and will not show the kind of gaps that Defense has used in the past to underscore U.S. defense needs."[109]

Weinberger and Casey agreed with Gates's recommendation to stop publication of the dollar comparisons. But this proved difficult, and soon problems materialized. SSCI asked for the defense spending comparisons, and a decision had to be made on permitting publication of an overall comparison of U.S. and Soviet GNP (including defense) for the JEC. SOVA began meeting with JEC staff about the upcoming DDI's testimony in November 1984.[110] Pressed hard by Senator Proxmire, Gates agreed to give the comparisons to the JEC. SOVA soon told Gates, however, that the JEC request was only one of many pending, including requests from the Department of Commerce, Supreme Allied Command Europe (SACEUR), the Office of Management and Budget (OMB), the Deputy for Defense Research and Engineering (OSD/DDR&E), the Director of Net Assessment in the Office of the Secretary of Defense (OSD), Strategic Air Command (SAC), and SSCI. A policy on release was needed, especially because some customers suspected that the decision to stop publication was based at least partly on what the trends showed.[111] Requests from DoD continued to pour in. DIA sent a memorandum to the DDI asking for more detailed dollar comparisons. While understanding his "discomfort" with the dollar estimates, DIA said "unfortunately, no acceptable substitute has been found for the measure of the size, direction, and significance of the overall Soviet military effort."[112]

SOVA Director Douglas MacEachin recommended in January 1985 that publication of the dollar comparison be resumed, arguing that the past several months had shown that customers would create their own estimates and that CIA needed to again be in position to control the quality of the comparisons and state the proper caveats.[113] Gates reluctantly agreed. As he wrote to Deputy National Security Adviser Robert MacFarlane, "Recent developments have made it necessary to reverse (my earlier) decision."[114] "I wanted to apprise you of these developments because of the earlier interest shown by you and John Poindexter." But Gates promised changes in the dollar-comparison paper, including a beginning section comparing accumulated physical stocks of military hardware, inserting in the key judgments section of the paper warnings of the limitations and proper uses of the comparisons, and relying as much as possible on percentage rather than absolute comparisons. At the DoD, Weinberger was briefed by Andrew Marshall on the preliminary results of the latest CIA dollar comparison. He told Weinberger that the most important trend showed U.S. procurement overtaking Soviet procurement in 1983 and that DoD should be prepared to respond.[115] On the positive side, Marshall said it was unlikely that the USSR would return to earlier rates of growth of defense spending for any sustained period, giving the United States an opportunity to challenge the Soviets in a way they would be hard pressed to match.

During the late 1970s and early 1980s demand increased for an expanded definition of national security spending, on the grounds that the traditional definitions used for calculating U.S. and Soviet defense outlays did not capture the entire national security-related burden. Gates asked SOVA to research the question so that the dollar-comparison paper could include a "broader view of the Soviet burden."[116] This effort, described in the next chapter, increased the Soviet burden relative to that of the United States—measured in their respective currencies. The ratio of the cost of Soviet national security programs to comparable U.S. programs, measured in dollars, was marginally higher than the ratio reflecting the more traditional definition of defense.[117]

The minor relief that comparisons based on an expanded definition of defense provided to defenders of the U.S. defense budget was offset by the results of an expansion in another direction—a comparison of the dollar cost of NATO and WP defense programs[118] which is discussed further in chapter 5. When DoD asked the agency for permission to give the finding to the JEC on an unclassified basis, DDI Richard Kerr demurred, saying he was reluctant to issue the paper in an unclassified form, and "in a form that will almost certainly be used against us."[119] SOVA replied that the agency should avoid being accused of selectivity (having already released the results of the U.S.-Soviet comparison to the JEC). In any case, "our dollar estimates have long been both in demand and controversial."[120] In reviewing the classified version of the paper, Gates, now DDCI, again voiced his concerns. Although he had few specific problems, he said it contained the same old caveats, which would be ignored: "One more time, our approach is an analytical construct, not a reflection of the real world." He predicted that the findings showing that the cumulative dollar costs of NATO defense programs in 1976–86 greatly exceeded the comparable costs for WP countries would be "precisely the part that will be leaked by the opponents of defense spending and used."[121] Clearly, high-level opposition to the dollar comparison was building again.

## Intracommunity Disputes over Economic Constraints on Soviet Military Programs

Through most of the 1970s and 1980s an underlying tension existed between the analysis of the Soviet economy and intelligence community expectations about the development of the USSR's armed forces. A study prepared for the JEC in 1977 summarized the agency's views on the outlook for the economy: reduced growth with no easy options to prevent the slowdown.[122] Still, the implications for defense were hardly drastic, according to the paper: "The slowdown in economic growth could trigger intense debate in Moscow over the future levels and pattern of military expenditures. Military programs enjoy great momentum and powerful political and bureaucratic support. We expect defense spending to continue to increase in the next few years at something like the recent annual rates of 4 to 5 percent because of programs in train. As the economy slows, however, ways to reduce the growth of defense expenditures could become increasingly pressing for some elements of the Soviet leadership."[123] The message that the Soviet economy was in trouble, although not new, began to percolate in wider circles, causing some to wonder whether the economic difficulties could lead to greater Soviet accommodation in arms control and restraint in military programs generally.

A few years later an interoffice task force led by OSR tackled the question of what these military programs would look like in the 1980s. More than forty individual research projects contributed to a major report published in early 1981.[124] In the process OER argued that the paper's projections of defense programs for the 1980s did not take seriously enough the possibility that economic constraints would force slower growth or even a downturn in Soviet defense spending. The approved paper concluded that programs in progress implied continued growth in defense spending in real terms through the 1980s, although "political strains resulting from growing economic problems could lead them (the Soviets) to moderate the growth of spending, particularly late in the decade."[125] Alternative scenarios considered two versions of stepped-up military programs: one featuring strategic forces and other general purpose forces, and one in which spending was reduced. The paper warned, however, that there was no evidence that the USSR planned to cut military spending. All three alternative scenarios to the baseline projection were considered unlikely.

Undoubtedly, if the late-1970s CIA estimates of Soviet military outlays had shown the slowdown in spending beginning in 1975 that later estimates did, the analysis in *The Development of Soviet Military Power* would have proceeded somewhat differently, giving greater weight to the downside of the force projections. The 1983 SOVA paper announcing the slowdown, for example, thought it more likely that the Soviets would continue the trend of little or no growth in procurement "in the hope of strengthening the economy for a long-term military competition."[126] But the idea of economic constraints on Soviet military programs did not win many adherents in other parts of the intelligence community, partly because many thought CIA's analysis of the Soviet economy was too bleak. The PFIAB review of the agency's work on the economy was sparked by suspicions of this kind.[127] There was a desire not to underrate the Soviet Union's capacity to sustain the military competition with the West. Thus Rowen, NIC chair in the early Casey years, turned back draft JEC testimony prepared in SOVA and ordered that it begin with a listing of basic Soviet economic strengths.[128] Gates, as DDI, rejected a draft economic paper, saying, "The continuing litany of 2 percent growth for the 1980s has less and less credibility with me." What, he asked, is the potential for policy changes to improve the economy?[129]

In the NIE arena, the proposition that economic constraints would impede the implementation of ongoing military programs had even less weight. We have already discussed the one-sided debate over the strategic forces estimate (NIE 11-3/8-85). The wide-ranging survey of Soviet strategic force developments delivered to two senate defense subcommittees by Gates and Lawrence Gershwin, NIO for Strategic Programs, in June 1985 illustrates the official view in the early Gorbachev years. They told the senators that the projected Soviet forces would "result in a growth of total Soviet strategic force expenditures of between 5 and 7 percent a year over the next five years," and 7–10 percent a year if Moscow decided to deploy ABM defenses on a wide scale.[130] Soviet economic problems, while severe, would not compel the USSR to give up important strategic programs or make significant concessions in arms control, although "the stark economic realities might cause some deployment programs to be stretched out." (The disparity between the flat spending on strategic programs of the recent past and the vigorous growth they projected for the future was not addressed.) Rather than using arms control to relieve economic pressures, the Soviet leadership would orchestrate its arms control policy to advance its "strategy of achieving strategic advantage" by limiting U.S. force modernization and blocking the Strategic Defense Initiative (SDI).

When Gorbachev announced his plans for industrial modernization, SOVA pointed out the conflict between his program and Soviet military objectives, especially in the machine-building and metal-working (MBMW) branch of industry, which was the source of most defense hardware and the machinery and equipment required for investment programs.[131] Indeed, as we noted in chapter 5, Gorbachev evidently planned to step up the growth of defense spending in the 1986–90 plan period. Also, the high-technology branches of MBMW (electronics, computers, communications equipment) were being counted on to upgrade the stock of fixed capital in the economy and close the technological gap between western and Soviet weaponry. A March 1986 SOVA paper described this civil-military competition.[132] It said the near-term effects on defense programs would probably not be large but that a crunch could develop in the late 1980s or early 1990s because Gorbachev's chances of success were dim: "Unless Gorbachev's efforts to modernize industry pay off in greater numbers of more advanced, high-quality equipment and in substantially increased productivity, the battle between civilian and defense interests will become more severe."[133]

The Washington policy community's reaction to this message was mixed. The paper was briefed to Secretary of State George Schultz and

Weinberger in the spring of 1986. Schultz was taken with the idea of an impending crunch and what it portended for Soviet interest in arms control.[134] Weinberger and his aides, on the other hand, thought the Soviet industrial modernization program, after a few years, would put the USSR in a better position to carry out its force modernization plans.[135] Even when it became clear that Gorbachev's economic program was foundering, there was generally great reluctance—whether within the intelligence community or in policy circles—to accept the idea that economic constraints would impinge seriously on Soviet defense programs.

In this vein, Gates, then acting DCI, asked in 1987 that a memorandum on the state of the economy focusing on Soviet strengths be sent to the secretary of defense.[136] The lead sentence of the resulting paper declared, "The Soviet economy, despite some ugly warts that make its operation costly and inefficient, is vested with great crude strength from the enormous resource base, and remains a viable system, capable of producing large quantities of goods and services annually, especially for industrial and military applications." The memorandum reported that even with no growth in GNP the Soviet Union could still spend the equivalent of $275 billion on its military. But zero growth was unlikely, and—although a relaxation of international tensions would help the modernization program "by freeing some resources that would otherwise be spent on defense"— Moscow had "the resources and the will" to increase defense outlays if the international climate did not improve. Industrial modernization and gains in living standards would be held down, "but it would not cause economic collapse or social upheaval."

## The End Game

As the Soviet Union disintegrated, the support and rationale for CIA estimates of the cost of Soviet military programs evaporated. First to go were the dollar comparisons. In August 1990 a SOVA memorandum considered the pros and cons of continuing the dollar estimates and recommended their continuance while putting consumers on notice that they would be eliminated in one to two years.[137] By then SOVA had dissolved its Defense Comparisons Branch and put the analysts to work on other accounts. Shortly thereafter, SOVA provided talking points to DDI John Helgerson to justify dropping the dollar estimates.[138] Marshall in mid-1991 urged that the dollar database be maintained because it "has been a major foundation for my office's work on a military investment balance report that top defense officials have found very useful in characterizing broad trends in our security situation."[139] In his reply to Marshall, Helgerson reaffirmed his decision to discontinue the dollar comparisons.[140] Letters to this effect were sent to the principal customers of the estimates on April 1, 1992.

Meanwhile the effort devoted to estimating the ruble cost of Soviet defense programs was scaled back to free analysts to work on problems connected with the unprecedented changes in the USSR. The comprehensive SCAM database was mothballed in favor of a spreadsheet model requiring much less data and operating at a higher level of aggregation. The costing contracts were also dropped, and the agency's commitment to defense costing was reduced from forty-five analysts, managers, and on-site contractors in the early 1980s to fourteen in early 1992. CIA management wanted the ruble work to continue, but at a much lower level of effort. Ironically more than a decade of refining the Soviet defense spending estimates was marked by increasing official doubts and disavowal. Still, the post-1975 research developed an improved set of estimates of the cost of Soviet defense programs. We present these in the next chapter.

# Chapter 5
# The Spending Estimates

After more than thirty years of development with advances and setbacks, CIA's estimates of Soviet military spending had by the late 1980s matured as much as they would, given the intelligence community's attention shift to the political turmoil in the Soviet Union. In this chapter we combine CIA's estimates of Soviet force levels and costs as of 1989–90 with the results of a 1980 review of the historical record to create a set of spending estimates covering the period 1951–90. The estimates are presented by resource category and military mission and reflect the broad changes in Soviet military programs and the shifts in defense priority during the Cold War.

It must be noted, however, that the pre-1965 estimates are based less solidly than the estimates for later years. Since the 1970s, CIA's practice had been to begin its annual reexamination of the physical evidence of the USSR's military programs first with 1960 and then with 1965. In 1980 a major review was undertaken to bring the physical data for the 1950s and early 1960s up to date. This effort resulted in a ruble estimate labeled "Historical Best" covering the years 1951–79 and expressed in 1970 rubles. To obtain the ruble series presented in this book, the estimates covering 1951–64 in the 1980 look back were converted to 1982 rubles using 1982/1970 ruble conversion factors, and were then linked with SOVA's 1990 estimate for 1965–90 in 1982 rubles.

The dollar estimates for 1951–64 are even shakier. As we noted earlier, the estimates of order-of-battle and military procurement for a given year tend to improve in accuracy as they recede into history and are viewed with the benefit of hindsight. Unfortunately the historical review carried out in

1980 did not produce a dollar estimate, so estimates of military outlays for 1951–64 in 1988 dollars shown here were obtained indirectly. The earliest year for which "good" spending data expressed in both 1982 rubles and 1988 dollars was available is 1965. These data were then used to derive 1982 ruble to 1988 dollar ratios for each resource category of spending at the mission-element level—for example, the estimate of 1982 ruble spending for aircraft procurement for the long-range air element of strategic attack forces in 1965 was divided by the counterpart 1988 dollar estimate for 1965, to obtain an implicit 1982 ruble to 1988 dollar ratio. These implicit ratios were then applied to the 1982 ruble estimates for 1951–64 to obtain the 1988 dollar estimates for those years. The use of these constant ratios for the entire period undoubtedly introduced error into the dollar estimates because the composition of spending within each category would not have remained constant over time.[1]

## The Question of the Price Base

The estimates in this section are compiled in 1982 ruble prices and 1988 dollar prices. The price base, of course, makes a difference for the level and composition of defense outlays. To indicate the extent of these differences, figure 5.1 shows Soviet defense outlays for 1965–79 (excluding RDT&E) in 1970 ruble prices, the 1982 ruble prices first used by CIA, and the 1982 ruble prices adopted after extensive research in the late 1980s for new prices in military equipment.[2] Although changes in estimates of force levels and production influence the levels and rates of growth of defense spending to some extent, the different price bases account for most differences in the estimates.

Inspection of the alternative estimates in figure 5.1 indicates that changing the price base does not significantly affect the trend, but the substitution of 1982 prices for 1970 prices has a major impact on the level of estimated Soviet defense spending. In fact, spending averages 72 percent higher in 1982 prices than in 1970 prices over the period 1965–79. Because estimates compiled in the late 1980s represent the best appreciation of Soviet force levels and production over time, they qualify as the "best" CIA estimate. In addition, the significant expansion of the price base (both rubles and dollars) that occurred in the 1980s provided a better foundation for costing the "physicals."

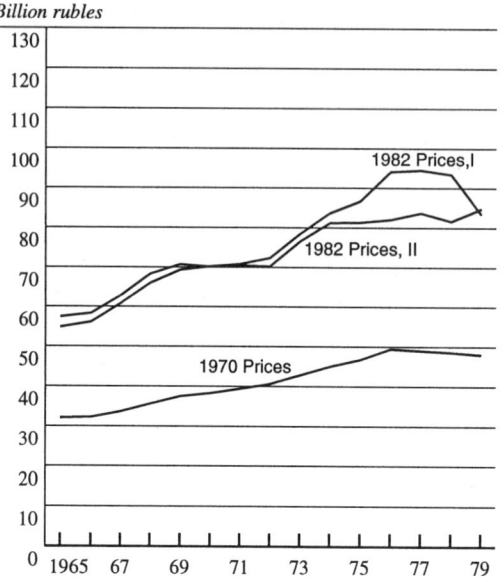

**Figure 5.1
Soviet Defense Outlays, 1965–79:
Alternative Price Bases** [a]

[a] Excluding RDT&E.

## Trends in Total Spending

Soviet spending on defense, broadly defined, was marked by periods of relative stability, rapid growth, and decline over 1951–90.[3] Marking off these periods is somewhat arbitrary, but the division into six periods depicted in table 5.1 and figure 5.2 seems reasonable. (Because the RDT&E estimate is probably less reliable than the other estimates—especially for year-to-year changes—total spending less RDT&E is also shown in the figure.) Viewed from this perspective, real Soviet military expenditures—measured in constant prices and therefore net of the effects of inflation—were essentially flat during most of the 1950s (1951–59) before beginning their most rapid increase (1960–69) of the forty-year period. Thereafter the rate of growth in outlays falls off (1970–74) and then decelerates even more (1975–84). For three years in the mid-1980s (1985–87) outlays climb sharply but drop just as abruptly in the last three full years of Soviet rule (1988–90). The trend in

TABLE 5.1

USSR REAL DEFENSE SPENDING BY RESOURCE CATEGORY, 1951–90[a]

| | | 1952–59 | 1960–69 | 1970–74 | 1975–84 | 1985–87 | 1988–90 |
|---|---|---|---|---|---|---|---|
| Procurement | Trend | *Down slightly* (-1.2% p.a.) | *Increases Sharply* (6.0% p.a.) | *Growth Subsides* (2.5% p.a.) | *Spending Levels Off* (-0.2% p.a.) | *Jumps Abruptly* (5.0% p.a.) | *Drops Sharply* (-7.6% p.a.) |
| | Major changes in spending | Sharp fall in aircraft production; other elements generally up, especially missiles | Missiles, surface ships and submarines, aircraft | Aircraft procurement increases; other elements increase slightly or decline | Substantial growth in land arms, space-related purchases offset by reductions in ships and subs, aircraft, and especially missiles | Led by purchases of aircraft and missiles | Especially aircraft, land arms, and space related equipment |
| Construction | Trend | *Big jump* (22.5% p.a.) | *Down substantially* (-3.8% p.a.) | *Dips somewhat* (-0.8% p.a.) | *Slow growth* (1.9% p.a.) | *Some growth* (2.3% p.a.) | *Spending levels off* (0.2% p.a.) |
| | Major changes in spending | Missile and general support facilities, airfields | Reduced construction of general support facilities | | | Mainly missile facilities | |
| Personnel | Trend | *Down sharply* (-5.5% p.a.) | *Up substantially* (3.0% p.a.) | *Continues to rise* (2.4% p.a.) | *Up slightly* (1.0% p.a.) | *Up slightly* (0.7% p.a.) | *Down substantially* (-2.6% p.a.) |
| | Major changes in spending | Pay and allowances fall as force levels decline | Post-Khrushchev buildup of general purpose forces | | Mainly due to rise in pension payments | | |
| O&M | Trend | *Down slightly* (-1.6% p.a.) | *Rapid growth* (11.0% p.a.) | *Strong growth* (5.4% p.a.) | *Slower growth* (2.4% p.a.) | *Substantial growth* (3.6% p.a.) | *Down substantially* (-2.6% p.a.) |
| | Major changes in spending | Decline in land arms and ship maintenance, utilities, POL | Increased costs of maintaining space facilities, ships, and missiles | Large increases in weapons maintenance, utilities costs, and pre-induction training | Continued increase in outlays for maintaining weapons and space-related gear | Expense of maintaining growing stocks of weapons and equipment | Weapons maintenance tails off |
| RDT&E | Trend | *Grows rapidly* (10.4% p.a.) | *Rapid growth continued* (11.2% p.a.) | *Growth subsides* (6.1% p.a.) | *Continued deceleration* (4.5% p.a.) | *Growth turns upward* (6.0% p.a.) | *Spending levels off* (-0.1% p.a.) |

[a] Based on constant 1982 ruble prices

**Figure 5.2**
**Soviet Defense Spending, 1951–90**

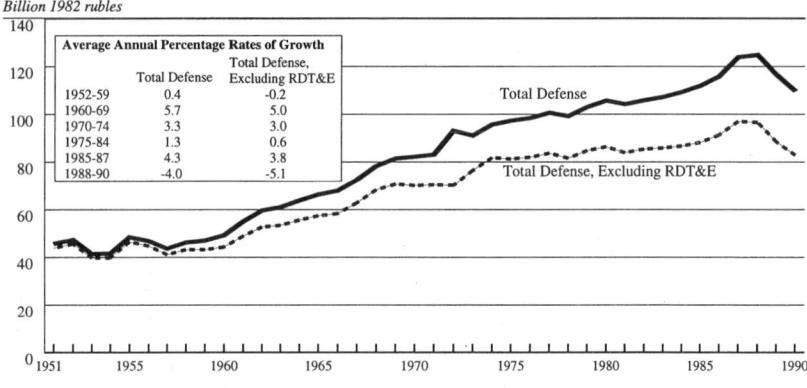

real Soviet military spending as shown in CIA's last thorough review of the physical evidence is quite different from what is probably the generally held sense of persistent, uniformly rapid upward movement since the 1950s. The sources of the variations in the estimated growth of total defense outlays must be sought at a more disaggregated level—in spending by resource category and by military mission.

## Spending by Resource Category and Military Mission

To understand the shifting patterns in Soviet military programs as they are reflected in the spending estimates, we will proceed using the periodization established above.

### Trends in Spending by Resource Category

We will first review the changes in the pace of programs over time for the major resource categories and then look at the changes in the distribution of real defense outlays among the resource categories. Table 5.1, which summarizes the behavior of these categories in 1951–90, serves as a convenient point of reference.

PROCUREMENT. The level of Soviet defense procurement as measured by CIA's estimate increased by 63 percent between 1951 and 1990. But the growth was far from even. Procurement trended downward through most of the 1950s and experienced ten years of rapid growth in the 1960s. Then the rate of growth fell off through the mid-1970s before procurement plateaued for ten years (1975–84). Under Gorbachev, procurement enjoyed three years of vigorous growth and then three years of even faster decline.

As would be expected, the volume of procurement was driven largely by varying rates of procurement of aircraft, missiles, major surface ships, submarines, and—to an increasing extent over the years—space-related equipment. These elements accounted for 80 percent of the procurement ruble in 1951, 75 percent in 1970, and 72 percent in 1990. The variations in procurement growth rates were partly due to the fact that successive generations of weapons systems entered into production at an uneven rate. Each generation of new hardware tended to be more expensive than its predecessor, but procured in smaller quantities than the generation it replaced.

Procurement was also shaped by major shifts in military policy, such as Khrushchev's coronation of the missile forces at the expense of the ground forces. Generally, any shift in the priorities given to various missions, whether initiated at what Stephen Meyer terms level one (Politburo/Defense Council) or level two (Ministry of Defense/General Staff), probably affected procurement growth because the missions differed in the shares of total spending they allotted to procurement.[4] Under these circumstances it is not surprising that the level of procurement grew unevenly and sometimes declined. The difficult and still unanswered question is whether the prolonged plateau in procurement from the mid–1970s to the mid–1980s and the spurt in procurement in 1985–87 were dictated by broader policy considerations.

The attention paid to CIA's estimates of procurement was substantial, and every evidence of small acceleration or decline in the rate of growth was probably taken more seriously than it should have been, at least by some. When CIA finally announced the leveling off in procurement begun in the mid–1970s, the discussion was heated and confused. As in earlier decades, critics of CIA's estimates asked, "How can procurement be growing so slowly when all around us we see evidence of a substantial buildup of Soviet military forces?" The confusion resided in the failure to recognize the distinction between stocks of military equipment and the flow of military equipment (the annual additions to the stocks). As long as initial procurement levels were high and the Soviets followed a practice of holding on to obsolescent equipment, stocks could increase rapidly even if the annual level of procurement remained the same or declined.

## Spending Estimates

CONSTRUCTION. The other part of military investment—construction—grew rapidly in the 1950s (more than 22 percent per year) but stayed at roughly the same level in real terms through the following three decades. The 1950s jump was fueled by the building of missile facilities, airfields, and other general support facilities. Construction of underground command and control and leadership protection facilities figured heavily in the crash programs that marked the decade.

PERSONNEL. The sharp reduction in the size of the armed forces initiated in 1953 came to an end in 1961. During the interval, outlays for personnel were cut by more than 25 percent. By 1975, with the reemphasis on the general purpose forces, personnel costs had climbed back to the 1953 level. Thereafter they grew slowly until Gorbachev ordered a force reduction in 1988.

Within the personnel cost category, pay and allowances in real terms increased less than personnel costs as a whole. Its share of total personnel costs fell from 63 percent in 1951 to 42 percent in 1990. In constant prices, pay and allowances can only increase with the numerical sizes of the armed forces and changes in the rank and position structure. Meanwhile improvement in the quantity and quality of the food ration pushed its share of personnel costs from 11 percent in 1951 to 19 percent in 1990. But the increasing burden of pension payments to military retirees proved to be the fastest-growing element of personnel costs. By 1990 its share amounted to 24 percent, whereas in 1951 it had been only 4 percent.

OPERATIONS AND MAINTENANCE. Over the entire period, outlays for O&M increased proportionately more than any other category of expenditures except for RDT&E. Like the costs of procurement, construction, and personnel, O&M costs declined in the 1950s. But in the 1960s and the first part of the 1970s, O&M outlays soared. Even in the period after the mid-1970s (and the 1988–90 partial demobilization), real O&M spending continued to rise at a healthy clip.

The rapid growth in O&M outlays was dictated by the pervasive expansion of inventories of weapons, equipment, buildings, and other physical holdings. These stocks had to be maintained, and maintenance became more expensive as the equipment grew more sophisticated. Indeed the increase in O&M outlays would have been even greater if some O&M were not driven by the "stock" of personnel. Because the size of the armed forces declined and then grew relatively slowly over the long term, the

O&M of personnel-related facilities (barracks and the like) did not increase as much as other elements.

RDT&E. CIA's estimates of Soviet military spending for RDT&E increased almost sixteen times—more than spending for any other category between 1951 and 1990. The rate of growth was particularly high in the 1950s and 1960s when it averaged about 11 percent per year. Although in the 1970s and most of the 1980s the rate of increase fell to roughly half of what it had been, RDT&E was still the fastest growing of all the resource categories.

## *Shifts in the Distribution of Spending by Resource Category*

Over the period 1951–90 the USSR spent the most (in 1982 prices) on military procurement. RDT&E, O&M, and personnel accounts were given roughly the same amount, but much less than procurement. Construction and civilian personnel trailed far behind in the allocations by resource category (table 5.2). Altogether the Soviets devoted about 3.3 trillion rubles (in 1982 prices) to defense in 1951–90, an amount equivalent to more than 16 percent of CIA's cumulative estimate of Soviet GNP (in 1982 established prices) in these years.

As figure 5.3 demonstrates, RDT&E, procurement, and O&M accounted for almost all of the growth in CIA's estimates of real Soviet defense spending. Outlays for construction, personnel (military), and civilian personnel increased less when spending for each period was converted to an average annual basis. The figure also illustrates the answer to the paradox cited above: how could the Soviet military buildup of weapons continue when the rates of growth in spending for procurement had leveled off? Clearly the substantial growth in procurement in 1970–74 meant that average annual procurement outlays in 1975–84 had to be higher than in the preceding five years, even if procurement did not rise after 1975. In other words the level of spending achieved by a spurt in spending in one period carried over into the following period.

The shares allocated to the different accounts did change substantially, with RDT&E and to a lesser extent O&M winning larger portions of the defense budget at the expense of procurement and personnel (figure 5.4). As the figure indicates, the shifts in the composition of spending were sustained and quite gradual except for the sharp drop in the share given to personnel after the 1950s.

Spending Estimates

TABLE 5.2
USSR DISTRIBUTION OF CUMULATIVE DEFENSE SPENDING
BY RESOURCE CATEGORY, 1951–90

|  | Billion 1982 rubles | Percentage share |
| --- | --- | --- |
| RDT&E | 522 | 16 |
| Procurement | 1,544 | 48 |
| Construction | 197 | 6 |
| Personnel | 415 | 13 |
| O&M | 515 | 16 |
| Civilian personnel | 44 | 1 |
| Total | 3,236 | 100 |

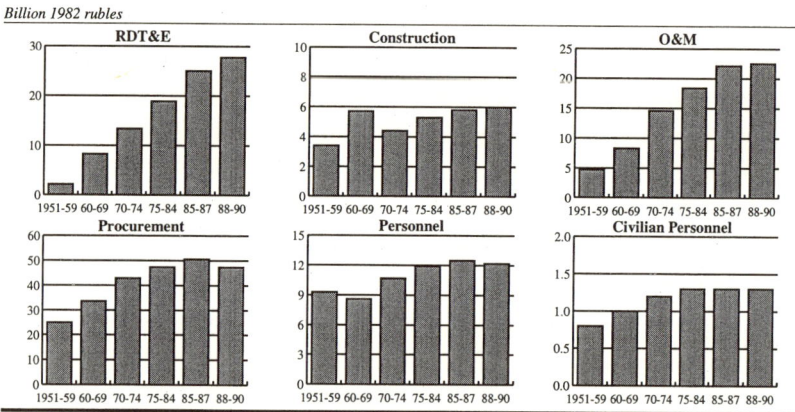

Figure 5.3
USSR: Average Annual Spending by Resource Category, 1951–90

## Trends in Spending by Military Mission

Sorting out spending by military mission permits an overview of how Soviet priorities changed during the Cold War.[5] Again, we will first review the trends in spending within each mission (shown in table 5.3) and then look at how each mission fared in terms of its share of total spending. Unfortunately, many expenditures—notably RDT&E, command and control, and intelligence and large outlays for general support—could not be distributed by mission.

Soviet Defense Spending

**Figure 5.4**
**Distribution of Defense Spending by Resource Category, 1951–90**
(Percentage Shares in 1982 Rubles)

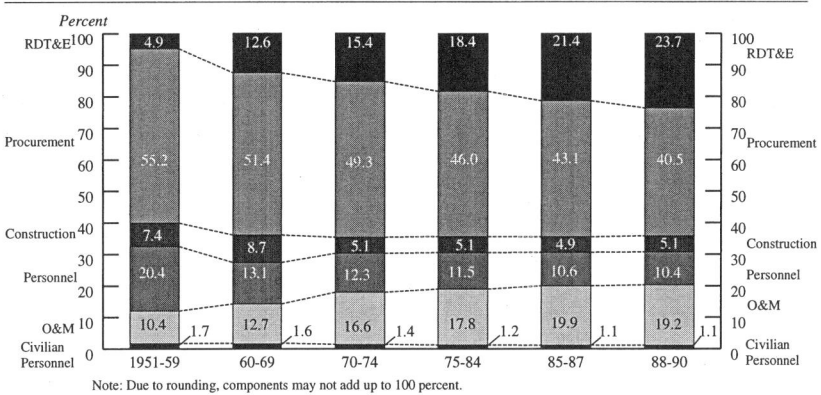

STRATEGIC OFFENSE. Since the 1950s, the West's attention has focused on the development of Soviet strategic offense. From a small beginning in the early 1950s, outlays for this mission soared. The rate of increase dropped off in the 1960s and again in the early 1970s. After 1975 expenditures on strategic offense trended downward.

In the 1950s fighter bombers and medium-range missiles intended for peripheral attack were responsible for the growth in mission outlays. By the 1960s ICBMs had taken the lead, sharing it with SLBMs (including submarines) in the 1970s. Bomber units designed for peripheral attack became another source of increased expenditures in the late 1970s and 1980s.

STRATEGIC DEFENSE. This mission experienced two periods of rapid growth: 1952–59 and 1985–87. In both instances a faster rate of augmentation of fighter-interceptor and surface-to-air (SAM) units accounted for the increase. During the other intervals outlays for fighter-interceptors and SAMs were fairly steady. Other mission elements played a smaller part in strategic defense costs.

GROUND FORCES. After a substantial fall in the mid- to late 1950s caused by large reductions in military manpower, outlays on ground forces grew steadily until the late 1980s. When personnel outlays leveled off, rising procurement of land arms and vehicles provided the growth.

Spending Estimates

TABLE 5.3

USSR REAL DEFENSE SPENDING BY MISSION, 1952-90 (1982 RUBLES)

| | | 1952-59 | 1960-69 | 1970-74 | 1975-84 | 1985-87 | 1988-90 |
|---|---|---|---|---|---|---|---|
| Strategic offense | Trend | *Spending soars* | *Continued rapid growth* | *Spending slows* | *Off slightly* | *Up slightly* | *Down substantially* |
| | Major changes in spending | (16.2% p.a.) Concentrated on peripheral attack missile units and long-range bomber units | (7.0% p.a.) Paced by spending on ICBMs and SLBM units; outlays for long-range bomber and MRBM elements fall | (3.8% p.a.) SLBM and ICBM units account for growth; spending on fighter/bombers down | (-2.3% p.a.) Outlays for ICBM and SLBM units down; increased spending on fighter/bomber elements | (1.4% p.a.) Led by rise in spending on long-range bomber units, outlays on IRBM units decline | (-5.2% p.a.) Especially outlays on SLBM and fighter/bomber units; spending on ICBM elements increases |
| Strategic defense | Trend | *Rapid growth* | *Growth decelerates* | *Spending falls* | *Down slightly* | *Rapid growth* | *Down sharply* |
| | Major changes in spending | (11.2% p.a.) Mainly interceptor aircraft and SAM elements | (2.8% p.a.) SAM and interceptor units still major factors in growth | (-3.4% p.a.) Outlays on SAM, interceptor, and ABM units turn down | (-0.9% p.a.) Decline in provision for interceptor units offset by growth elsewhere | (8.2% p.a.) Spurred by increased spending for interceptor and SAM elements | (-4.5% p.a.) Outlays for interceptor and SAM units down; spending on ABM elements up |
| Ground forces | Trend | *Down sharply* | *Resurgence* | *Growth slows* | *Up slightly* | *Some acceleration* | *Outlays plummet* |
| | Major changes in spending | (-5.1% p.a.) | (4.5% p.a.) | (2.5% p.a.) | (1.4% p.a.) | (3.3% p.a.) | (-8.0% p.a.) |

| | | | | | | | | |
|---|---|---|---|---|---|---|---|---|
| Nonstrategic navy | Trend | *Big cut* (-6.8% p.a.) | *Spending rebounds* (1.7% p.a.) | *Outlays fall* (-2.0% p.a.) | *Up slightly* (1.8% p.a.) | *Some acceleration* (3.2% p.a.) | *Spending levels off* (-0.4% p.a.) |
| | Major changes in spending | Huge decline in naval air; major surface ships pared back while submarine force increases | Spending on submarine units increases | Spending on submarines and naval auxiliaries down | Paced by increases in spending on submarine and carrier units | Broad-based, submarine, carrier, and major surface ship elements | Increases on outlays on carrier aircraft and submarine units offset by decline in carrier and major and minor surface ship elements |
| Tactical air | Trend | *Huge decline* (-15.3% p.a.) | *Rebound* (5.8% p.a.) | *Spending soars* (15.8% p.a.) | *Levels off* (-0.6% p.a.) | *Outlays turn up* (2.3% p.a.) | *Sharp decline* (-7.1% p.a.) |
| | Major changes in spending | Mainly in light bomber units, but also fighters | Focused on fighters, reconnaissance, and transport units | Again, fighter and transport units, but also light bombers | Increases in fighter and reconnaissance units counterbalanced by fall in spending on light bomber units and drones | Sparked by spending on reconnaissance elements | Mainly due to fall in spending on light bomber and fighter units |
| Other | Trend | *Substantial growth* (6.0% p.a.) | *Accelerated growth* (12.6% p.a.) | *Rapid growth* (5.3% p.a.) | *Slower growth* (2.9% p.a.) | *Acceleration* (5.0% p.a.) | *Substantial decline* (-3.0% p.a.) |
| | Major changes in spending | Primarily due to RDT&E | Big increase in RDT&E, intelligence and communications | RDT&E, intelligence and communications, and general support | Continued increases in RDT&E, intelligence and communications | Especially RDT&E, intelligence and communications, and airlift | Drop in spending on intelligence and communications, air, lift, and general support |

NAVY. Through most of the Cold War the conventional naval mission received relatively level funding. It was cut back substantially in the 1950s, largely because of a steep decline in naval air units. Thereafter overall allocations to the big-ticket items—major surface ships, submarines, and naval air elements—moved within a narrow range. When the rate of growth varied, it was usually because of an acceleration or deceleration of spending on submarines, although after 1975 increased outlays for aircraft carriers and surface ships helped underwrite some expansion.

TACTICAL AIR. The swings in spending on tactical aviation were greater than in other missions. The large-scale demobilization of light bomber units in the 1950s resulted in a steep decline in outlays. Spending rebounded in the 1960s and even more in the early 1970s, with increased allocations to fighter-bombers, transports, and reconnaissance aircraft. From the mid-1970s to the mid-1980s expenditures leveled off and then turned up; they declined in the late 1980s because of the changes noted in table 5.3.

OTHER OUTLAYS. This catchall collection of activities grew persistently and sometimes rapidly (in 1960–69) because of the sustained growth in RDT&E and intelligence and communications.

### Shifts in the Distribution of Spending by Mission

Over the forty years covered in the CIA estimates, two-fifths of real spending was allocated to the "other" outlays. The naval and ground force missions followed with one-sixth each, and the tactical air mission was given about one-twelfth (table 5.4). In terms of absolute growth, period by period, the "other" category (led by RDT&E, intelligence and communications, and general support) exhibited steady, substantial gains (figure 5.5). Spending on strategic offense increased rapidly but peaked in 1975–84, while outlays for strategic defense were relatively flat over all periods. Ground forces outlays grew in the 1970s and 1980s, but those for the conventional naval mission declined somewhat in the first three periods before increasing slightly in the last three. The tactical air mission lost ground in the 1960s and then recovered briskly in the 1970s and early 1980s.

The shares of total outlays devoted to each mission are shown in figure 5.6. Because of the rapid growth in the portion given over to "other" activities—from 15 percent in 1951–59 to 50 percent in 1988–90—the share devoted to the rest of the missions taken together dropped dramatically.

TABLE 5.4
USSR DISTRIBUTION OF CUMULATIVE
DEFENSE SPENDING BY MISSION, 1951–90[a]

|  | Billion 1982 rubles | Percentage share |
|---|---|---|
| Strategic offense | 349 | 11 |
| intercontinental attack | (225) | (7) |
| peripheral attack | (124) | (4) |
| Strategic defense | 322 | 10 |
| Ground forces | 517 | 16 |
| Navy | 522 | 16 |
| Tactical air | 246 | 8 |
| Other | 1,279 | 40 |
| Total | 3,234 | 100 |

[a]Totals and percentages are based on unrounded data and may not equal the sum or percent of the rounded components shown in the table.

Figure 5.5
USSR: Average Annual Spending by Mission, 1951–90

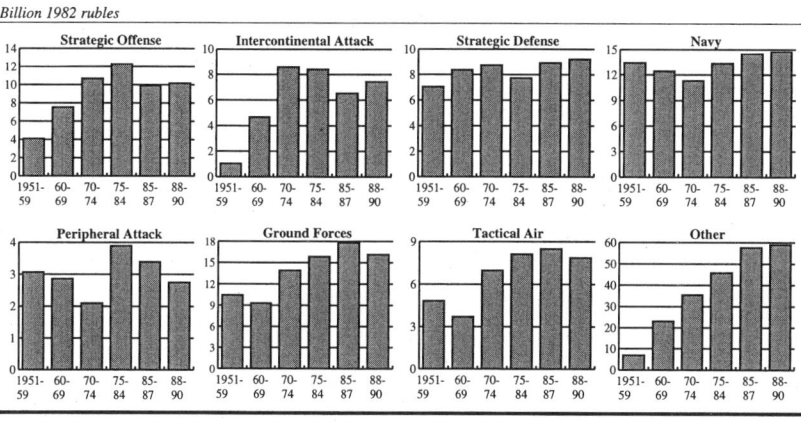

The intercontinental attack mission, however, basically held its own after an early rise to prominence. The naval mission was the big loser, as its share declined by 16 percentage points. The ground forces and strategic defense ranked next, with shares reduced by 7–8 percentage points, while the peripheral attack and tactical air missions each suffered an erosion in shares of 4 percentage points.

Spending Estimates

**Figure 5.6**
**Distribution of Defense Spending by Mission, 1951–90**
**(Percentage Shares in 1982 Rubles)**

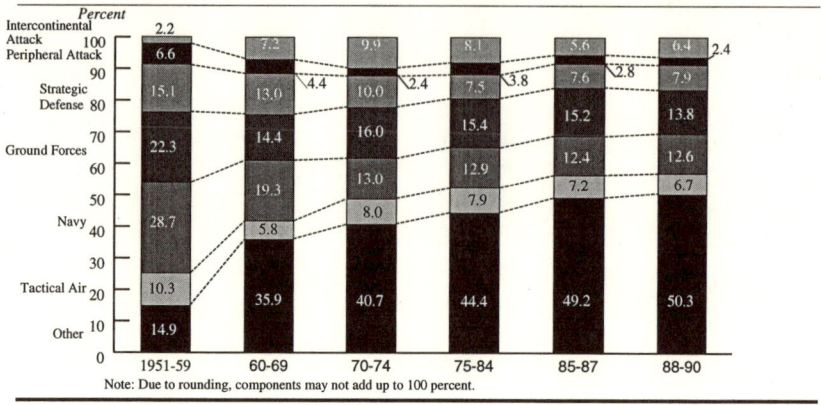

Note: Due to rounding, components may not add up to 100 percent.

## Comparisons with U.S. Defense Spending

Because the price base for the comparisons of Soviet and U.S. defense programs changed with every new estimate update, we cannot present a comparison in a single set of prices. Instead, we present one comparison for 1951–64 and another for 1965–90. Because the last comparison for the earlier period in a single set of dollar prices was compiled in 1966, it does not represent the most recent assessment of the physical dimensions of Soviet military programs in the pre-1965 period. (As noted earlier, the look back in 1980, in an effort to reflect the latest physical estimates for the pre-1965 period, unfortunately dealt only with the ruble estimates, not the dollar estimates.) It does show what CIA was telling policymakers in those years, however. The comparisons are dollar comparisons, not an average of comparisons carried out in both dollar and ruble prices. We consider critics' objections to this procedure in the next chapter.

### The 1951–64 Period

During 1951–64 CIA's estimates of the dollar cost of Soviet defense programs were $65 billion less than U.S. outlays for national defense (in 1972 prices).[6] After the Korean War buildup and subsequent build down, U.S. outlays were relatively stable in constant prices (figure 5.7), although they climbed from $39 billion in 1955 to $50 billion in 1964 in current prices.

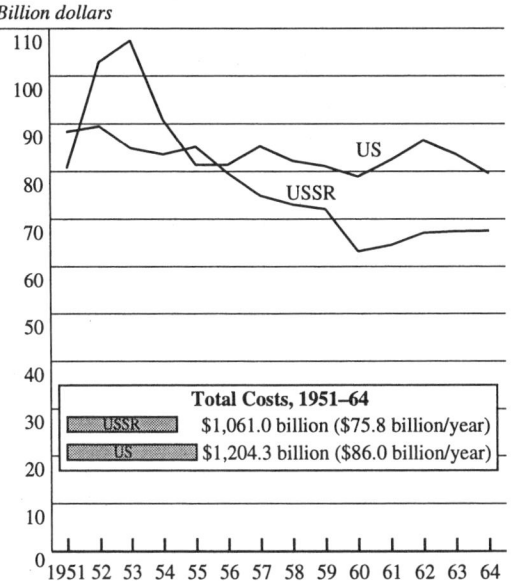

**Figure 5.7**
Dollar Cost of U.S. and Soviet Military Programs, 1951–64 (Billion 1972 Dollars)

The dollar-equivalent costs of Soviet programs reflect the Khrushchevian demobilization of the late 1950s, which was arrested after 1960.

Distributions of USSR dollar costs by mission and resource category are available for all years of the 1961–64 period, but in different dollar price bases. Comparisons of dollar costs of U.S. and Soviet programs could be found for only 1960 and 1964, and only by resource category for 1960. (The DoD did not compile mission breakdowns of its budgets until the 1960s.) These are reported in table 5.5.

Not surprisingly, the USSR's numerical advantage in the size of the armed forces translates into a large dollar difference when U.S. rates of pay and allowances are applied to Soviet personnel. The U.S. programs were substantially larger, however, in the other resource categories—procurement, construction, and O&M—according to these early CIA estimates.

Spending Estimates

TABLE 5.5

## COMPARISONS OF U.S. DEFENSE OUTLAYS AND DOLLAR VALUATIONS OF SOVIET MILITARY PROGRAMS, 1960 AND 1964[a]

| By resource category | 1960[b] (billion 1959 dollars) | | |
|---|---|---|---|
| | United States | USSR | USSR as Percentage of United States |
| Personnel | 11 | 17 | 155 |
| Procurement | 12 | 10 | 83 |
| O&M | 11 | 6 | 55 |
| Other | 10 | 8 | 80 |
| Total | 44 | 41 | 93 |
| By resource category | 1964[c] (billion 1972 dollars) | | |
| | United States | USSR | USSR as Percentage of United States |
| RTD&E[c] | 12.9 | 7.0 | 54 |
| Nuclear programs | 1.4 | 1.6 | 118 |
| Procurement | 22.8 | 15.8 | 69 |
| Construction | 1.8 | 1.0 | 55 |
| Personnel | 19.5 | 29.2 | 150 |
| O&M | 18.3 | 11.5 | 63 |
| Total | 76.7 | 66.2 | 86 |
| By misson | United States | USSR | USSR as Percentage of United States |
| Strategic attack | 7.0 | 6.0 | 87 |
| Strategic defense | 1.7 | 5.2 | 308 |
| General purpose forces | 16.5 | 14.1 | 85 |
| RTD&E[d] | 13.6 | 7.6 | 56 |
| Command and general support | 37.9 | 33.2 | 88 |
| Total | 76.7 | 66.2 | 86 |

[a] Totals and percentages are based on unrounded data and may not equal the sum or percent of the rounded components shown in the table.
[b] CIA, Office of Research and Reports, *Soviet Military Expenditures, 1958–65*, CIA/RR ER SC 61-4, April 28, 1961, p. 28.
[c] CIA/DI, *Soviet Spending for Defense: An Annual Review*, vol. 2, *A Monetary Comparison of Soviet and U.S. Defense Activity*, SR IR 73-12, August 1973, pp. 49–50, 57–58.
[d] In the distribution of dollar costs by resource category, the costs of active military personnel engaged in RDT&E are included in personnel costs; in the breakdown by mission, the costs of military RDT&E personnel are recorded under the RDT&E mission.

Nuclear programs, broken out of procurement in the 1964 distribution, were an exception.

U.S. and Soviet differences in the pattern of mission effort were already well established in the 1950s and early 1960s. The Soviet effort in strategic defense was far larger, while U.S. emphasis on tactical aviation accounts for the U.S. margin over the Soviets in general purpose forces. In strategic defense, the Soviet programs were especially large relative to their U.S. counterparts in SAMs and fighter-interceptors. This was a reflection of the larger potential threat U.S. bombers posed to the USSR, as well as the huge landmass Soviet air defense forces had to protect from the U.S. intercontinental threat and NATO forces in Europe.

## *The 1965–89 Period*

According to CIA's dollar comparisons for 1965–89 (compiled in 1988 dollars in August 1990) the cost of U.S. programs continued to exceed the equivalent dollar cost of Soviet programs until 1972 (figure 5.8), when the impact of the Vietnam War on U.S. spending began to subside. From 1972 through 1985 the USSR surged ahead—in some years by a wide margin. By the mid-1980s the costs of the U.S. and Soviet programs were about equal. In all, the United States spent $6.1 trillion (1988 prices) cumulatively in 1965–89 while the dollar-equivalent cost of Soviet programs was an estimated $6.3 trillion.

The mission allocation of these dollar costs differed greatly, however. As figure 5.9 indicates, the cost of strategic offense and strategic defense was far greater for the USSR than for the United States. Likewise the dollar-equivalent cost of Soviet ground forces was substantially more than U.S. spending on this mission. On the other hand, the United States allocated more to the tactical air, naval, and mobility missions.[7] In addition the United States provided more in the way of support for its missions. (The cost of RDT&E is excluded from this comparison.)

The distributions of dollar costs of U.S. and Soviet military programs by resource category are more similar than the distributions by mission (table 5.6). The major differences in spending by resource category are that personnel costs are higher for the USSR than the United States, reflecting higher manpower levels, while the cost of O&M is much higher in the United States than in the USSR over the twenty-five-year period. The influence of the Vietnam War on U.S. spending can be seen clearly (figure 5.10), as can the defense buildup beginning late in the Carter administra-

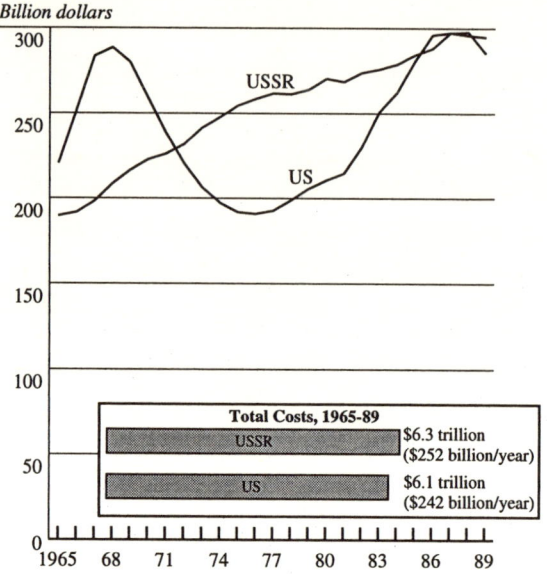

**Figure 5.8**
Dollar Cost of U.S. and Soviet Military Programs, 1965–89 (1988 Dollars)

tion. The pattern of the dollar cost of Soviet military programs was evident earlier in the discussion of ruble costs—steady growth in O&M and RDT&E, the procurement and personnel plateau beginning in the mid-1970s, and almost level costs for construction.

## The NATO-Warsaw Pact Excursion

The Soviet-U.S. comparisons were sometimes criticized as misleading because of their failure to take into account the contribution of military allies on both sides. Requests from Congress and the Pentagon for a NATO-WP comparison increased in the 1970s. OSR resisted making such comparisons, arguing that developing dollar costs for each NATO and WP country with the same degree of rigor devoted to the U.S.-Soviet comparisons would be too burdensome relative to the benefits gained.[8] A few years later the Working Group on Soviet Military Economic Analysis chaired by Selin

**Figure 5.9**
Dollar Cost of Soviet and U.S. Military Programs by Mission, 1965–89

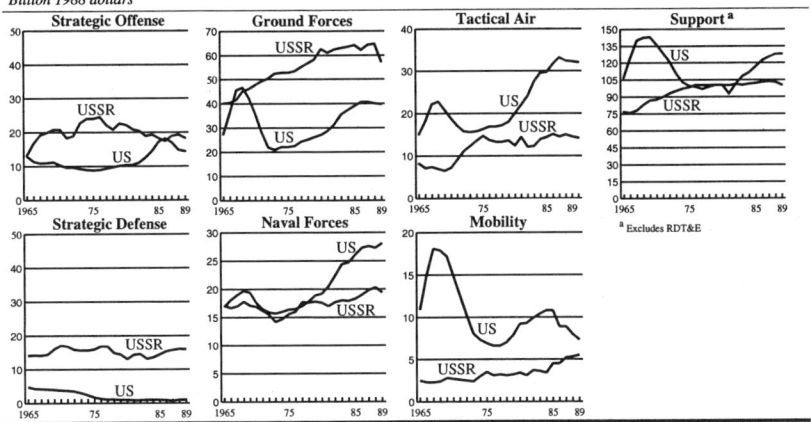

**Figure 5.10**
Dollar Cost of Soviet and U.S. Military Programs by Resource Category, 1965–89

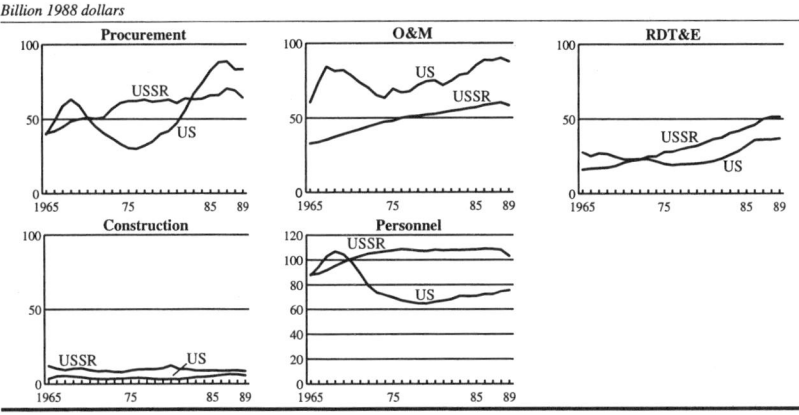

found a "widespread desire" within the intelligence community and among congressional staff members for NATO-WP comparisons.[9] The group recommended that in the longer term a program "should be initiated to concentrate on NATO and Warsaw Pact forces," although the effort was considered lower priority than some of the group's other recommendations.[10]

In following up on the working group's report, SOVA turned to the estimation of the dollar-equivalent costs of the defense programs of the

Spending Estimates

## TABLE 5.6
## UNITED STATES AND USSR: DISTRIBUTION OF DEFENSE SPENDING IN TERMS OF DOLLARS, 1965–89

| By mission[a] | billion 1988 dollars | | percentage | |
| --- | --- | --- | --- | --- |
| | United States | USSR | United States | USSR |
| Strategic offense | 291 | 502 | 4.9 | 8.0 |
| Strategic defense | 53 | 362 | 0.9 | 5.8 |
| Ground forces | 793 | 1,369 | 13.3 | 21.8 |
| Navy | 503 | 433 | 8.5 | 6.9 |
| Tactical air | 560 | 298 | 9.4 | 4.7 |
| Mobility forces | 130 | 84 | 2.2 | 1.3 |
| Support | 2,852 | 2,378 | 48.0 | 37.8 |
| Other[b] | 765 | 868 | 12.9 | 13.8 |
| Total | 5,947 | 6,294 | 100.1 | 100.1 |

| By resource category | billion 1988 dollars | | percentage | |
| --- | --- | --- | --- | --- |
| | United States | USSR | United States | USSR |
| Procurement | 1,355 | 1,456 | 22.8 | 23.1 |
| Construction | 106 | 237 | 1.8 | 3.8 |
| O&M | 1,899 | 1,218 | 31.9 | 19.4 |
| Personnel | 1,947 | 2,599 | 32.7 | 41.3 |
| RDT&E | 639 | 783 | 10.8 | 12.4 |
| Total | 5,946 | 6,293 | 100.0 | 100.0 |

[a]Totals and percentages are based on unrounded data and may not equal the sum or percent of the rounded components shown in the table.

[b]Includes RDT&E and outlays for the strategic mission that cannot be allocated to either strategic offense or strategic defense—i.e., those for production of nuclear materials and for command and control systems.

non-Soviet Warsaw Pact (NSWP) members. An April 1986 report (*Non-Soviet Warsaw Pact Defense Costs: 1970–84*, SOV 86-10019 CX) found little growth in NSWP costs, whether measured in domestic currencies or dollars. Late the following year SOVA and the Office of European Analysis published a NATO-WP comparison after completing research to establish the purchasing power dollar equivalents of non-U.S. NATO defense budgets (CIA/DI, *A Comparison of Warsaw Pact and NATO Defense Activities, 1976–86*).

The comparisons' results are set out in tables 5.7 and 5.8. In dollar terms, NSWP outlays were almost flat in 1976–86, and non-U.S. NATO spending grew slowly. But over the period the dollar cost of non-U.S. NATO defense programs was almost three times greater than the dollar cost of NSWP programs. As a result the Soviet "advantage" over the United States in the dollar cost of military programs was overturned when the comparison was expanded to include U.S. and USSR allies. NATO outlays exceeded WP outlays in every year of the period and by an increasing margin in the 1980s. Altogether the dollar cost of NATO military programs in 1976–86 amounted to $3.5 trillion compared with $3.1 trillion for the WP.

When the paper's findings were briefed a year before its publication, they provoked a good deal of skepticism, if not outright opposition. John J. Maresca, the Deputy Assistant Secretary of Defense for European and NATO policy, feared that the analysis could "lead to a distorted view that NATO is outspending the Warsaw Pact and getting less for its investment."[11] He asked that the study address the qualitative differences in the equipment procured by NATO and WP as well as the differences in operating practices. To meet these objectives, the paper was revised to include an extensive discussion of these issues.

Within CIA DDCI Gates reiterated his basic, longstanding concerns about dollar costing. He argued that the data on cumulative spending by

TABLE 5.7

VALUE OF NON-SOVIET WARSAW PACT AND NON-US NATO DEFENSE ACTIVITIES, 1976–86[a]

|  | Billion 1985 Dollars | | Percentage Shares | |
| --- | --- | --- | --- | --- |
|  | NSWP | Non-US NATO | NSWP | Non-US NATO |
| *Total* | 499 | 1,401 | 100 | 100 |
| *Investment* | 109 | 237 | 22 | 17 |
| Procurement | 88 | 189 | 18 | 13 |
| Construction | 22 | 48 | 4 | 3 |
| *Operating* | 373 | 1,091 | 75 | 78 |
| Personnel | 219 | 566 | 44 | 40 |
| O&M | 154 | 525 | 31 | 37 |
| RDT&E | 17 | 73 | 3 | 5 |

[a]Totals, subtotals, and percentages are based on unrounded data and, therefore, may not equal the sum or percent of the rounded components shown in the table.

Spending Estimates

TABLE 5.8
DOLLAR VALUE OF WARSAW PACT AND
NATO DEFENSE ACTIVITIES BY RESOURCE CATEGORY, 1976–86[a]

(billion 1985 dollars)

| | 1976 | 1977 | 1978 | 1979 | 1980 | 1981 | 1982 | 1983 | 1984 | 1985 | 1986 | Total |
|---|---|---|---|---|---|---|---|---|---|---|---|---|
| Warsaw Pact total | 263 | 265 | 269 | 274 | 279 | 280 | 281 | 286 | 289 | 293 | 297 | 3,075 |
| Investment | 82 | 82 | 82 | 85 | 85 | 85 | 82 | 83 | 83 | 84 | 84 | 914 |
| procurement | 70 | 70 | 70 | 72 | 72 | 71 | 69 | 70 | 70 | 72 | 72 | 778 |
| construction | 12 | 12 | 12 | 12 | 13 | 12 | 13 | 13 | 13 | 13 | 12 | 137 |
| Operating | 153 | 154 | 156 | 158 | 161 | 163 | 165 | 167 | 168 | 169 | 171 | 1,785 |
| personnel | 92 | 93 | 94 | 95 | 97 | 97 | 98 | 98 | 99 | 98 | 99 | 1,059 |
| O&M | 61 | 62 | 62 | 64 | 64 | 66 | 67 | 68 | 69 | 71 | 73 | 726 |
| RDT&E | 28 | 28 | 30 | 31 | 33 | 34 | 35 | 37 | 38 | 40 | 42 | 375 |
| NATO total | 276 | 282 | 287 | 294 | 302 | 313 | 330 | 349 | 361 | 373 | 378 | 3,544 |
| Investment | 49 | 51 | 54 | 62 | 65 | 72 | 81 | 92 | 100 | 106 | 109 | 842 |
| procurement | 41 | 44 | 47 | 55 | 58 | 65 | 73 | 83 | 91 | 97 | 100 | 753 |
| construction | 8 | 8 | 7 | 7 | 7 | 7 | 8 | 9 | 9 | 9 | 9 | 89 |
| Operating | 202 | 205 | 207 | 205 | 209 | 213 | 220 | 225 | 226 | 229 | 230 | 2,370 |
| personnel | 98 | 98 | 96 | 96 | 97 | 98 | 100 | 102 | 102 | 103 | 105 | 1,094 |
| O&M | 104 | 107 | 111 | 110 | 112 | 115 | 120 | 123 | 123 | 126 | 125 | 1,276 |
| RDT&E | 25 | 26 | 26 | 27 | 27 | 28 | 29 | 32 | 35 | 38 | 39 | 3321 |

[a]Totals and subtotals are based on unrounded data and may not equal the sum of the rounded components shown in the table.

the two alliances "will be leaked by the opponents of (U.S.) defense spending and used."[12] In response MacEachin, the director of Soviet Analysis, pushed for publication, maintaining that the dollar estimates "do have utility" and noting that customers in DoD and Congress had been asking for a NATO-WP comparison. He wrote that "we think it would be intellectually dishonest to not publish our findings when we have already reported on the spending gap favoring the USSR when only the United States and the Soviet Union are compared."[13]

In the next two years Gorbachev announced a reduction in Soviet military spending and the WP disintegrated, removing the need to carry out additional NATO-WP comparisons. The comparisons published at the end of 1987 provided a useful alternative perspective on the East-West military competition, but one that probably should have been researched a decade earlier. In particular, they highlighted a neglected fact in discussions of the East-West military balance—that the NATO partners of the United States had a larger defense effort than the Soviet Union's WP allies.

## Consistency of the Spending Estimates with Internal and External Developments Affecting the Soviet Union

The trends we have sketched in Soviet military spending can be given a credibility check by seeing if they are generally consistent with internal and external shifts in Soviet policy and circumstances. Because the issues are so complex and the sources of Soviet behavior during the Cold War are still being debated, the check must be preliminary and impressionistic. In table 5.9 we note in chronological order some of the milestones in Soviet internal politics and military decision-making as well as the external developments that may have affected decisions on military spending. The calendar begins in 1960 quite simply because CIA estimates of the course of spending in the 1950s have not provoked as much contention as those covering the post-1960 period. In broad terms we looked for evidence that would reject the CIA estimate that Soviet defense spending increased rapidly in the 1960s, leveled off in the early 1970s, accelerated briefly in the mid-1970s, and then grew slowly until 1986, when there was another short-lived acceleration. (These trend changes were depicted in figure 5.2, which represents CIA's estimate of Soviet spending in 1982 rubles, excluding RDT&E.)[14]

The 1960s began with Khrushchev still in the midst of the general force reduction he announced in January 1960. Tensions with the new

TABLE 5.9

LANDMARKS IN THE SOVIET MILITARY BUILDUP

|  | Internal Developments | Military Developments | Foreign Policy Developments |
|---|---|---|---|
|  | <<1960–69: Defense Spending Increases Sharply in Real Terms>> | | |
| 1959(?)–60 |  | Khrushchev's force reductions |  |
| Jan 60 |  | Khrushchev announces 1/3 cut in military manpower |  |
| 1961 |  | Khrushchev cancels force reduction, raises announced defense budget by one-third | Berlin crisis |
| Fall 62 |  |  | Cuban missile crisis |
| Oct 64 | Khruschev deposed; Brezhnev succeeds him |  | China sets off atomic device |
| Feb 65 |  |  | Kosygin mission to China fails |
| Summer 65 |  | Consensus forms for all-service military buildup |  |
| Fall 65 |  | Soviet buildup vs. China begins |  |
| Sep 65 | Kosygin outlines economic reform package |  |  |
| Dec 65 |  |  | Mao kicks off cultural revolution and breaks ties with CPSU |
| Post-65 | Ideologues gain in party |  |  |
| Apr 67 |  | Defense Council meets with Soviet commanders |  |
| June 67 |  |  | Arab-Israeli Six-Day War |
| Jun 67 |  |  | Kosygin-Johnson meeting |
| Spring-Summer 68 |  |  | Czech crisis |
| Oct 68 |  | WP invasion of Czechoslovakia |  |
| Jul 68 |  |  | Non-proliferation agreement |
| Mar 69 |  | Sino-Soviet border clashes begin |  |
| Spring-Summer 70 |  |  |  |
| Aug 70 |  |  | Soviets deploy air defense troops to Egypt; Brandt, Brezhnev sign agreement |

## TABLE 5.9 CONTINUED

|  | Internal Developments | Military Developments | Foreign Policy Developments |
|---|---|---|---|
|  |  |  | recognizing existing frontiers of Poland, FRG, GDR |
| Nov-Dec 71 |  |  | India-Pakistan War; threat of Chinese intervention |
| Mar-Apr 71 |  | Brezhnev outlines "Peace Program" |  |
|  | <<1970–74: Growth in Real Defense Spending Subsides>> | | |
| May 72 |  |  | Nixon visits USSR; agreements on offensive missiles, Joint US-USSR commercial commission; |
| Oct 72 |  |  | Soviet-US trade agreement |
| May 73 | Nat'l security triumvirate added to Politburo |  |  |
| Jun 73 |  |  | Brezhnev visits US; signing of agreement on prevention of nuclear war |
| Oct 73 |  |  | Yom-Kippur War—Soviets threaten to intervene |
| Jul 74 |  |  | Brezhnev-Nixon sign agreements on nuclear arms control |
| Aug 74 Oct 74 | Brezhnev launches Farm Program |  |  |
| Oct 74 |  |  | Brezhnev-Kissinger secret deal over Jewish emigration; |
| Nov 74 |  |  | Brezhnev-Ford talks in Vladivostok |
| Dec 74 |  |  | Soviets repudiate emigration commitment |
|  | <<1975–84: Real Defense Spending Levels Off>> | | |
| Jan 75 |  |  | USSR abrogates Trade Agreement in response to US |

TABLE 5.9 CONTINUED

| | Internal Developments | Military Developments | Foreign Policy Developments |
|---|---|---|---|
| | | | disclosure of emigration agreement & passage of Jackson-Vanik Amendment |
| Mid-1970s | | Soviets deploy SS-20s in EE and Far East | |
| Winter 75 | | | Cuba intervenes in Angola with Soviet help |
| Apr 75 | 10th Five Year Plan cuts planned investment growth in half | | |
| Jan 77 | | Brezhnev Tula speech renounces pursuit of nuclear superiority | |
| Fall 77 | | | Soviet-Cuban intervention in Somali-Ethiopian War |
| May 78 | | USSR begins to fortify Northern Territory | |
| Aug 78 Dec 78 Jan 79 | | | Japan signs Friendship Treaty with China, Vietnam occupies Cambodia |
| Jun 79 | | SALT II signed | |
| Jul 79 | | | |
| Dec 79 | Brezhnev delivers controversial speech to CC plenum | | |
| Dec 79 | | USSR invades Afghanistan | |
| Aug-Sep 80 | | | Gierek falls; Solidarity gains |
| Feb-Mar 81 | XXVII Party Conference | | |
| Jul 81 | | Ogarkov article warns of war | |
| Nov 81 | | Warning net activated | |
| Dec 81 | | | Martial law in Poland |
| May 82 | New Brezhnev farm program | | |
| Jul 82 | | | Strong attack on Reagan foreign policy by Arbatov |
| Oct 82 | | Brezhnev tells military leaders to make better use of resources | |
| Nov 82 | Brezhnev dies, replaced by Andropov | | |
| Mar 83 | | | Reagan's "Evil Empire" speech; announces SDI |
| Sep 83 Nov 83 | | | Andropov declares USSR forced to |

## TABLE 5.9 CONTINUED

| | Internal Developments | Military Developments | Foreign Policy Developments |
|---|---|---|---|
| | | | respond to Reagan's crusade vs socialism; Soviets back off START and INF talks |
| __84 | | Decision to adopt temporary defensive posture in West | |
| Sep 84 | | Orgarkov removed as Dep Min Defense and Chief of General Staff | |
| | <<1985–87: Acceleration in Growth of Real Military Spending>> | | |
| Mar 85 | Chernenko dies, replaced by Gorbachev | | |
| Nov 85 | | | Reagan, Gorbachev agree on: - inadmissibility of nuclear war -not to seek nuclear superiority -to try to prevent arms race in space - principle of 50% cut in ICBMs |
| Dec 85 | 1986–90 Plan reportedly called for 40% rise in military spending | | |
| Oct 86 | | | Reykjavik Summit—Soviets offer large concessions on ICBM & IRBM cuts |
| Dec 86 | | | Soviets tell Afghans of intention to withdraw |
| Jan 87 | Gorbachev spells out need for democratization at party plenum | | |
| Jun 87 | Economic reform package announced | | |
| | <<1988–90: Real Defense Spending Drops Abruptly>> | | |
| Mid-88 | | Political & military leaders conclude unilateral force cuts are feasible | |
| Dec 88 | | In UN speech Gorbachev announces cuts in defense spending and in forces | |

Kennedy administration—the 1961 Berlin crisis in particular—may well have helped overturn the decision on force reductions. In any event the Cuban missile crisis, Khrushchev's removal, and China's increasingly independent and unfriendly stance would have supported Soviet political and military leaders pressing for an all-around buildup of military forces. Marshal

**Spending Estimates**

Rotmistrov, for example, in a March 1965 *Kommunist* article, attacked the "subjectivism" inherent in Khrushchev's conviction that any major East-West clash would quickly become nuclear. Other writers outlined a spectrum of possibilities for conflict, including a prolonged conventional war.

Later in the decade, the 1967 Arab-Israeli War, the decision to suppress a reformist regime in Czechoslovakia, and skirmishes on the Soviet-Chinese border undermined the case of any leaders who might have wanted to restrain the growth in defense spending. At home, reform was in retreat as Prime Minister Kosygin's reform package was scaled back and ultimately discarded, the victim of internal opposition and its own inherent weaknesses.[15]

The picture in the 1970s is unclear. On the one hand, Soviet rates of economic growth were falling and the USSR was forced to import record amounts of foreign grain to sustain its consumer programs, especially Brezhnev's ambitious farm program announced in late 1974. In 1970 the USSR and the Federal Republic of Germany (FRG) signed an agreement recognizing the existing Polish, German Democratic Republic (GDR), and FRG frontiers. Early in 1971 General Secretary Brezhnev outlined his Peace Program. Holloway claims that Brezhnev's statement was the central feature of Soviet détente policy in the 1970s and concludes that it had its origins in the Sino-Soviet border clashes, President Nixon's successful exploitation of the Sino-Soviet quarrel, socialist ascendancy in the FRG, Soviet achievement of strategic parity with the United States at the end of the 1960s, and a desire to obtain foreign technology.[16] The following two years saw visits by President Nixon to the USSR and Brezhnev to the United States and the signing of agreements on strategic missiles, commercial cooperation, trade, and the prevention of nuclear war. The high point of détente diplomacy was perhaps reached at the June 1973 summit meeting. All of these developments were consistent with a Soviet decision to slow the growth of defense spending.

But the military spending estimate shows a sharp acceleration in 1973 and 1974, largely a result of increased outlays on tactical aviation. This disparity between what the Soviets were apparently doing and how western analysts described prevailing Soviet policy illustrates the difficulty of relating current spending to current policy. Military programs have a long gestation period, and while it is possible to curtail production when weapons have finished RDT&E, Brezhnev may have been unwilling or unable to make the decision as Gorbachev did later to suddenly scale back programs.

In any event the signals from the 1970s are conflicting. On the one hand, the United States and the USSR concluded a number of arms control and commercial agreements, and Brezhnev in a 1977 speech at Tula denied that the USSR was aiming for a first-strike capability. On the other hand, the Soviet-U.S. agreements soon dissolved in conflict over emigration and congressional actions. Some observers maintain that the military retained a dominant position in the leadership through the late 1970s.[17] The decision to scale back the planned growth in new fixed investment in the 1976–80 and 1981–85 economic plans has been ascribed to a desire to support defense growth and some gains in consumption in the face of slowing overall growth.[18] Yet the estimated slowdown in growth of military spending continued through the end of the 1970s. Perhaps the civilian and military leadership jointly concluded that defense outlays were already high enough to support the military establishment they needed.[19] In addition the defense-industrial base was probably plagued by the same productivity problems as the rest of Soviet industry. Meanwhile estimated defense outlays in current prices continued to rise, so Soviet leaders may have had no sense that they were restraining the real growth of defense programs.

Brezhnev seems to have been able to avert an acceleration in defense spending until his death. Indeed in May 1982 he pushed through another farm program that promised to appreciably increase the share of agriculture in investment and other products of industry and construction. But internal opposition was building.[20] The Soviet invasion of Afghanistan with its subsequent fallout in terms of boycotts and embargoes, the rise of the solidarity movement in Poland, and above all the sharp increase in U.S. military spending begun under President Carter reinforced the arguments of Soviet military leaders and their supporters that war with the West was still foreseeable.[21] According to a former high-ranking KGB officer, some sort of war scare took place in November 1981 as KGB and Soviet military representatives in NATO countries were ordered to report urgently on indicators of an imminent U.S. nuclear attack.[22] Meanwhile in October 1982, the month before he died, Brezhnev told an assemblage of military leaders that they needed to make better use of their resources, implying that their demands would be met to a lesser degree than usual.

When Yuri Andropov replaced Brezhnev in November 1982, the stage seemed set for an upswing in Soviet military spending. President Reagan's announcement of SDI in March 1983 should have increased the likelihood of such an outcome.[23] Andropov declared on September 28, 1983, just before he died, that because the United States was out to destroy socialism and achieve military domination, the USSR would have to respond to preserve the military balance.[24] The upswing in Soviet defense outlays did not take place until 1985, according to CIA's estimates. MccGwire argues that Soviet policy toward the United States was rethought in 1983 and 1984, leading to an acceptance of poor to indifferent relations with Washington but a desire to reduce tensions in Western Europe by adopting a defensive posture there.[25] At the same time the rigidities of a failing five-year economic plan and the inevitable time lag between initial discussion and decision and the actual outlay of resources would have prevented a visible, immediate upswing in military outlays.

At any rate the next five-year plan (1986–90), according to Gorbachev, called for a 40 percent rise in military spending.[26] Gorbachev would have participated in the decision to increase defense outlays substantially. Indeed his agreement to support higher spending may have been a condition for winning military backing in his bid to become general secretary in January 1985. He believed his policies could restore economic growth and thus permit higher defense budgets, higher living standards, and industrial modernization. He became convinced by mid-1987 that far-reaching economic reform and relief from military competition with the West was essential to maintain Soviet power. By mid-1988, Soviet political and military leaders had agreed (the military leaders with considerable reluctance) that unilateral cuts in Soviet forces were possible, and at a speech at the United Nations in December 1988 Gorbachev announced that such reductions would occur.

## The Burden of Defense

From the beginning of the Cold War, the West paid a great deal of attention to the Soviet defense burden. In a free market, competitive economy expenditures for defense programs would measure accurately the resources used to support a defense effort. At the same time, with some qualifications, the outlays on defense would represent the opportunity cost of defense programs—the value of the civilian goods and services foregone by the country's decision to spend on defense. Measuring the Soviet defense burden was not so simple.

### The Resource Cost of Soviet Military Programs

Beginning in the 1950s western observers agreed that Soviet military programs imposed large costs on the economy. Indeed in the early 1950s a principal question posed to U.S. economic intelligence was whether the USSR could support such a large military establishment. The defense resource burden was generally represented by the share of defense in Soviet GNP, and this number was prominent in almost every discussion of Soviet defense or economic trends.

The latest CIA estimates of the ratio of defense outlays to Soviet GNP are presented in table 5.10. Defense spending and GNP are both valued in 1982 ruble prices in one of the calculations and, where possible, in current

TABLE 5.10
RATIO OF SOVIET DEFENSE SPENDING
TO GNP (PERCENT)

|      | 1982 prices | Current prices |
|------|-------------|----------------|
| 1951 | 24.2        |                |
| 1952 | 23.4        |                |
| 1953 | 19.4        |                |
| 1954 | 18.3        |                |
| 1955 | 19.5        |                |
| 1956 | 17.1        |                |
| 1957 | 15.0        |                |
| 1958 | 14.3        |                |
| 1959 | 13.6        |                |
| 1960 | 14.5        |                |
| 1961 | 15.3        |                |
| 1962 | 16.1        |                |
| 1963 | 16.6        |                |
| 1964 | 15.7        |                |
| 1965 | 16.0        |                |
| 1966 | 15.6        |                |
| 1967 | 15.8        |                |
| 1968 | 16.1        |                |
| 1969 | 16.4        |                |
| 1970 | 15.4        | 12.2           |

Spending Estimates

## TABLE 5.10 CONTINUED

|      | 1982 prices | Current prices |
|------|-------------|----------------|
| 1971 | 15.1        |                |
| 1972 | 14.9        |                |
| 1973 | 15.2        |                |
| 1974 | 15.4        |                |
| 1975 | 15.5        |                |
| 1976 | 15.0        | 13.6           |
| 1977 | 15.0        |                |
| 1978 | 14.4        |                |
| 1979 | 15.0        | 14.7           |
| 1980 | 15.3        |                |
| 1981 | 14.9        | 14.4           |
| 1982 | 14.8        | 15.0           |
| 1983 | 14.6        | 14.4           |
| 1984 | 14.7        | 15.2           |
| 1985 | 14.9        | 15.9           |
| 1986 | 14.9        | 16.4           |
| 1987 | 15.7        | 17.6           |
| 1988 | 15.5        | 17.8           |
| 1989 | 14.3        | 15.6           |
| 1990 | 13.8        |                |

prices in the other. The definition of defense is broad—that is, it includes KGB and interior ministry troops, construction and railroad units, and the regular armed services. The share of resources going to defense is better measured by the ratio in current prices than the ratio in constant prices because the current price measure reflects changes in the relative productivity of resources used in the economy. Although the agency recognized that the share of defense in GNP in constant prices was a second-best measure, neither GNP nor defense was estimated in current prices until the late 1970s and the 1980s. Thus the gaps in the table.

Turning first to the ratio in constant 1982 prices, this measure of defense burden declines rapidly in the 1950s and early 1960s and then varies between 14 and 16 percent through 1990. It should be remembered, however, that the revision of the ruble estimates in 1976 (in 1970 prices) pushed the ratio up from 6 percent to 12–13 percent (in 1970 prices), leading many to believe that something dramatic was taking place in the Soviet economy; in fact the agency's underestimate of hardware and RDT&E costs meant that the defense burden was higher than estimated. The estimates of forces and activities in physical terms, past and present, had not changed.

The ratio of defense to GNP in current prices (insofar as it can be tracked) was somewhat different. It was lower than the constant price ratio in the 1970s and higher in the 1980s. As a result, according to these measures the burden of Soviet defense programs increased by almost 6 percentage points between 1970 and 1988 in current prices but essentially did not change in constant prices. This finding stems directly from CIA's measures of higher inflation in the defense sector compared with the economy generally.

Even if the measures of the resource cost of defense were accurate in Soviet established prices, they were flawed because of the weaknesses of Soviet pricing. Because prices were planned centrally rather than competitively, the ratio of the price of one good to the price of another generally did not reflect either the relative utility of the two goods to buyers or the relative quantity of resources used in their production.[27] CIA, following in the footsteps of economists at Columbia University, Harvard University, and the RAND Corporation, attempted to convert estimates of Soviet GNP and defense to a factor cost basis, which would provide a better measure of the resource content of the goods and services included in GNP. The procedure involved deducting indirect taxes and profits from the cost of products and adding back the amount of subsidies and a return on capital and land. This recalculation (in constant 1955, 1970, or 1982 prices) usually added 1 or 2 percentage points to the burden measure.

In the end, however, the effort to estimate defense at factor cost is a failure. The factor cost adjustments for fuel, food, and many other supplies delivered to the military could be made in a comparable manner in the civilian and military arenas. The Soviet conscript could be given the pay that his counterpart in industry was earning. But the degree of subsidization afforded to defense industry was never satisfactorily taken into account, nor was the cost of the special privileges given it. The factor cost adjustment procedure implicitly assumed that a defense enterprise in the machine-building branch or, say, the chemical industry earned the same rate of profit and was given the same degree of subsidization as the civilian

Spending Estimates

enterprises. To add to the muddle, it was not entirely clear whether the prices of military hardware that CIA had obtained were special prices charged to the military or full-cost prices (i.e., unsubsidized). If they were subsidized to a greater degree than other machinery prices, the agency's factor cost adjustment—and the measure of the burden—was too small. Such subsidization—even if quite large—would not have increased the defense share of GNP by more than 2 or 3 percentage points at most.

### Expanded Definition of Defense

CIA's burden measures were criticized on other grounds. One charge gathered momentum in the 1970s and the 1980s: even the agency's broad definition of defense did not capture all of the national security-related spending in the Soviet Union.[28] DCI Casey became interested in the "cost of empire" and asked for a quick survey, which led to ruble and dollar estimates of an expanded definition of defense.[29] CIA's first cut at what it called an extended comparison of U.S. and Soviet defense programs appeared in May 1986.[30] The activities covered in the expanded definition are shown in table 5.11, which compares 1983 U.S. and Soviet national security programs in 1984 dollars. The baseline definition of defense is the narrow definition previously employed in all dollar comparisons and thus excludes the internal security, railroad, and construction troops and outlays on civil defense. For the United States, the broader definition is 42 percent greater in dollar terms than the baseline definition; for the USSR, the broader definition is 55 percent greater. (In later estimates, the incorporation of military pensions in the baseline definition reduced these differences by about half for both countries.) In other words the change in definition increased the difference in the relative size of national security programs in the two countries from 12 percent in the USSR's favor to roughly 22 percent—a difference appreciably less than most cost-of-empire disciples probably anticipated.

An effort to find the ruble cost of an expanded definition of national security proceeded parallel with the work on the expanded dollar cost comparison. The results were summarized in a 1987 SOVA paper.[31] As figure 5.11 indicates, the inclusion of additional national security outlays added more than 3 percentage points to the defense burden in 1980 but less in other years. In this connection, the imputed value of Soviet economic aid fell in the 1980s as the opportunity cost of supplying energy to Moscow's client states declined because of the narrowing difference between world

## TABLE 5.11
## ESTIMATED COSTS OF SELECTED U.S. AND SOVIET NATIONAL SECURITY ACTIVITIES IN 1983[a]

| Activities | United States | USSR | USSR as percent of United States |
|---|---|---|---|
| | (billion 1984 dollars) | | |
| Traditional defense activities (baseline definition) | 208 | 232 | 112 |
| Other national security activities | 87.9 | 123.5 to 131.7 | 140 to 150 |
| *Mobilization/wartime preparedness* | 6.4 | 23.5 to 29.6 | 367 to 462 |
| internal security troops | 2.1 | 3.0 | |
| railroad and construction troops | 3.0 | 9.0 | |
| civil defense | .2 | 7.5 to 8.9 | |
| industrial and strategic reserves | .2 | 0.1 to 0.4 | |
| defense highways | 0 | 3.6 to 5.2 | |
| industrial surge capacity | 0.01 | 0.3 to 3.0 | |
| synthetic fuels | 0.4 | 0.04 | |
| merchant fleet O&M | 0.4 | negligible | |
| *Enhancement of global position* | 19.4 | 31.6 | 163 |
| economic aid | 3.0 | 11.3 | |
| military aid | 6.8 | 5.1 | |
| conduct of foreign affairs | 1.9 | 2.9 | |
| foreign information and exchanges | 0.6 | 4.0 | |
| government-funded foreign students | 0.1 | 1.5 | |
| civil space | 7.0 | 7.0 | |
| *Past defense activities* | 62.1 | 68.4 to 70.5 | 110 to 114 |
| veteran's benefits | 29.5 | 13.9 | |
| civilian pensions | 7.9 | 6.6 | |
| military pensions | 17.1 | 43.0 | |
| defense-related lawsuits | 0.1 | negligible | |
| cash flow (interest on national debt attributable to defense less new debt created for defense) | 7.6 | 4.9 to 7.0 | |
| Traditional defense plus other national security activities | 295.9 | 356 to 364 | 120 to 123 |

[a]Totals, subtotals, and percentages are based on unrounded data and may not equal the sum or percent of the rounded components shown in the table.

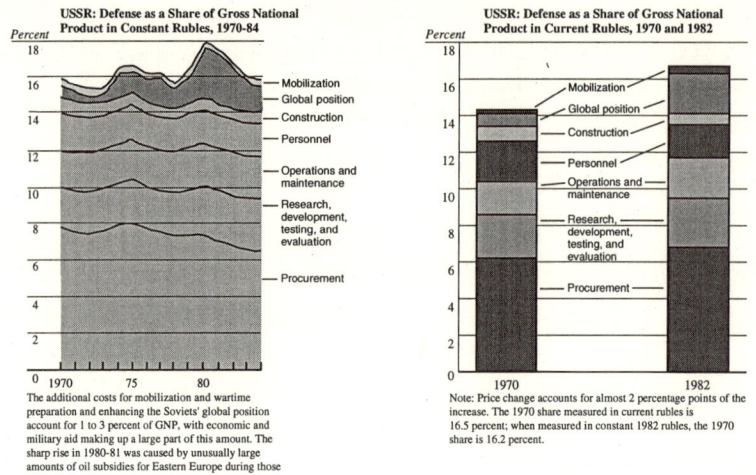

Figure 5.11
Defense as a Share of Soviet GNP Under Alternative Definitions of National Security

The additional costs for mobilization and wartime preparation and enhancing the Soviets' global position account for 1 to 3 percent of GNP, with economic and military aid making up a large part of this amount. The sharp rise in 1980-81 was caused by unusually large amounts of oil subsidies for Eastern Europe during those years.

Note: Price change accounts for almost 2 percentage points of the increase. The 1970 share measured in current rubles is 16.5 percent; when measured in constant 1982 rubles, the 1970 share is 16.2 percent.

market prices for energy and prices charged in intra-CEMA (Council for Economic Mutual Assistance) trade. For all the debate about what should be counted in defense costs, a thorough survey of possible additions did not seem to change the picture.

Still, the valuations of many of these activities was tricky and controversial. For example, a RAND study, *The Costs and Benefits of the Soviet Empire, 1981–1983*, estimated the ruble cost of empire (trade subsidies, economic aid, trade credits, military aid, Afghan War, and covert and related activities) at 36 billion rubles in 1982, whereas the SOVA estimate at the time was 17 billion rubles. Much of the difference was accounted for by differences in the way trade subsidies to Eastern Europe were estimated in dollars and then converted to rubles. RAND tried to calculate the cost to the USSR of paying higher than world prices for imports from Eastern Europe and accepting lower than world prices for its exports to Eastern Europe. SOVA dealt only with Soviet energy exports to Eastern Europe. Once calculated in dollars, the subsidies had to be converted to rubles—by the official exchange rate in SOVA's estimate and by an estimate of what the hard-currency portion would sell for in rubles in the USSR if embodied in imports. The problem was that the imports would sell for very different amounts of rubles depending on what they were—grain, machinery, or consumer goods.

Toward the end of the 1980s interest in the calculation of an expanded definition died down. CIA's 1986 estimate appeared in the 1988 dollar paper. By then the opportunity costs of Soviet exports to its client states in terms of foregone hard currency earnings had declined considerably, and the Soviet Union announced a unilateral cut in military spending. The controversy over what a true definition of the Soviet defense burden should include still smolders, but now it is an academic debate no longer driven by current policy interests.

## *Economic Impact of Defense Programs*

In assessing the economic impact of Soviet defense spending, CIA analysts tried to push beyond the traditional burden measure—the defense share of GNP. Uncomfortable with the estimate (pre-1976) that defense was only 6 percent of GNP, they pointed to the military's preferential access to scarce material resources, its advantages in hiring the best managers and skilled workers, and its preemption of the USSR's best research and development facilities. This formulation became standard boilerplate in most discussions of the Soviet defense burden.[32] It was a way of saying that the effect of the allocation of resources to military programs is greater than that implied by the share of defense in GNP in established prices or at factor cost.

The focus on the economic impact of defense programs was another way of getting at the question of the opportunity cost of defense outlays. The question was addressed at two levels. At the micro or industry level, a series of studies examined the defense claim on output, from basic sectors like steel and petroleum to the high-technology sectors of Soviet machine-building (electronics, robotics) and the chemical industry. In every instance the military demand was substantial and the higher the level of technology involved, the greater the military share of total output. Nonetheless it was hard to make the case that the defense establishment's demands were a larger constraint on the economy in the 1960s, 1970s, or 1980s than in the 1950s, given the overall growth in GNP or in the sectors supplying goods and services to the military. In fact, such microconstraints on production played a significant role in NIEs of the mid-1950s, only to fade from view thereafter except for discussions of scientific capabilities.

A second strain of analysis tried to model the effect of increases or reductions in the growth of real defense spending, a dynamic (as opposed to static) approach to the measurement of the opportunity costs of defense.[33] CIA's SOVSIM econometric model of the Soviet economy attempted to assess the effect of a change in the level of defense spending on GNP and nondefense GNP over time. The results of the modeling were perhaps not unexpected, but they were unconvincing to many. They indicated that small changes in the growth of defense spending produced even smaller changes in the growth of nondefense GNP and still smaller changes in the rate of growth of overall GNP. Not surprisingly the diversion of even a million military servicemen and -women to civilian pursuits could not radically affect the growth of a labor force of 155 million (in mid-1988). Alternatively a substantial reduction in military procurement could permit more investment and a faster growth of the Soviet capital stock. But the overall effect on economic growth was still not impressive, in part because the model reflected the relatively low return to increases in the capital stock that was characteristic of Soviet economic development from the 1950s on.

Again, the impact on the civilian economy of transferring resources from defense to the civil sector depended on whether these resources would be more productive, just as productive, or less productive than the labor, fixed capital, and land already employed in the civilian economy. Those who stressed the high quality of the resources engaged in the defense sector thought a transfer of resources out of defense would have a larger impact than economic models of the Soviet economy implied. But how much larger was anyone's guess. On the other hand, some argued that the effect of a defense cutback on civil production might be less than the usual analysis indicated. Rush V. Greenslade of OER suggested that the fault resided in the "gross and pervasive 'disequilibrium'" that characterized the Soviet economy.[34] Because of this disequilibrium, the opportunity cost of defense spending was not equal to the cost of the resources allocated to defense.[35] The reason for the indeterminate nature of opportunity costs in the Soviet military included existing wide differences in resource productivity among producing sectors and the difficulty a planned economy has in transferring resources from military to civilian uses. The advantages enjoyed by defense-industrial enterprises—priority access to supplies, minimal interference by local party chieftains, and a close connection with a demanding customer—would be absent in civilian production. On balance, Greenslade thought the increase in civilian output resulting from a transfer of resources from the military sector would be less than the cost of these resources.[36]

In sum, an analytically satisfactory definition of the Soviet defense burden eluded CIA's analysts throughout the Cold War. The distortions in Soviet prices and the inflexibility in Soviet planning obscured the true opportunity cost of the USSR's defense programs. Nonetheless a defense share of GNP that had climbed to 18 percent in current prices by the late 1980s (three times the comparable U.S. share) was certainly large enough to concern Soviet leaders.

## Influence on Allocations to Defense

CIA analysts in the 1950s sometimes maintained that defense programs were being curtailed because of competing civilian programs. From the 1960s to the early 1980s, however, NIEs and a range of CIA products tended to assert that, because of its high priority, defense outlays would continue to increase (within reasonable limits) despite growing difficulties in the economy. The 1982 finding that the growth of defense spending had slowed several years earlier touched off a debate over the reasons for the slowdown and the prospects for its continuing.

The basic outlines of a discussion repeated again and again in CIA and intelligence community publications appeared in NIE 11-4-54.[37] Briefly noting the competition of military production with investment and consumption, the estimate said that Soviet leaders, intent on accelerating the production of consumer goods, "apparently intend to limit defense outlays to approximately the high level reached in 1952 and maintained in 1953." By the early 1960s the description of economic constraints had altered. A NIE in 1963, for example, reiterated the already established analysis that civil-military competition for resources was most severe in the machinery sector and that defense enjoyed priority access to skilled manpower and high-quality materials.[38] Looking to the future, the estimate declared that the USSR could afford higher levels of defense spending but not without affecting programs to raise living standards and perhaps reducing the future rate of industrial growth. Soviet leaders, according to the estimate, seemed willing to risk these consequences although the "problem of resource allocation will continue to plague" them. Two years later, another estimate said that even if defense outlays increased by 20 percent in the new five-year plan, the Soviet economy could shoulder the burden and at the same time gradually improve the equipment and technology of Soviet industry and the standard of living.[39] It added, however, that defense programs would be "increasingly subjected" to reviews of their costs and effectiveness.

When Soviet economic growth slowed still more in the 1970s and CIA published a pessimistic appraisal of Soviet economic prospects, the view that military programs were not threatened remained entrenched in national intelligence and CIA publications. Thus a memorandum to holders of NIE 11-4-78 provided a grim description of the USSR's economic situation, including a "growing possibility of social instability," and reported that "if military spending continues to grow at 4 percent per year the increment in national output going to defense could rise from 1/5 to as much as 3/4," severely limiting the leadership's scope for maneuver. Still, the "projection of Soviet military spending most consistent with available evidence suggests that pressures in favor of continuing the existing arms buildup are likely to offset any inclination toward change that might arise from the leadership's growing economic concerns." If necessary, "appeals to a more extreme patriotism" and "repressive measures" would be used to increase production and maintain domestic control in face of mounting military budgets.[40]

The tenuous connection between defense burden measures and CIA's projections of Soviet defense programs was apparent in the agency's reaction to the abrupt 1976 increase in the estimate of Soviet defense outlays. Asked by Senator Charles Percy of the JEC how defense was impinging on the civilian economy and what effect growing consumer demands would have on an economy still spending so much on defense, CIA replied: "The new CIA estimates of Soviet defense spending—50 to 55 billion rubles in 1975—have altered significantly our perceptions of the economic costs of the Soviet defense effort. Analysis of the complex issues of economic burden and resource allocations is still in its preliminary stage. However, it is clear that the Soviets are far more willing than we thought to forgo growth in the civilian sector (and consumer satisfaction) in favor of expanding military capabilities." The response added that "there are no indications that the leadership has seriously considered diverting resources from military use in response to consumer demands." It noted that the Soviet population had traditionally recognized that programs to boost industrial production and defend the country justified "a slow growth in living standards."[41]

Developments in the Soviet Union in the 1980s and the process of the intelligence community's coming to terms with the deceleration in the estimated growth of Soviet defense spending restored the defense burden as a significant factor in projection of Soviet defense programs. In 1983–84 the discussion of the origins of the slowdown unearthed an array of possible causes, but they centered on adaptations to current difficulties in the economy while downplaying the possibility that a strategic decision had been made to limit the growth in the defense budget. Then the debate over Soviet defense policy heated up when Gorbachev became general secretary in 1985.

CIA analysts thought Gorbachev's ambitious investment program in support of his goal of modernizing the USSR's production base would "leave little or no room for increasing defense spending procurement above its recent rates."[42] In fact, as we have shown, defense spending accelerated in the first year of the 1986–90 plan and Gorbachev later said this growth had been planned. Gorbachev likely believed his focus on labor discipline and a shake-up of the party and managerial ranks would result in productivity gains that would permit substantial growth in defense procurement and the production of equipment for investment. CIA's analytical line at the time was that the capacity to produce the strategic weapons coming on line was already in place, suggesting "that Gorbachev can handle the defense/growth tradeoff for the time being."[43] The real test for Gorbachev would be when demands to expand and renovate defense industries were renewed: "If the modernization program is not going well, Gorbachev will have to deal with military leaders asking for more defense-related investment and advocates of devoting even more resources to modernization—here the plan begins to unravel and choices have to be made."

The reforms in the economy introduced in mid-1987 proved disruptive rather than corrective. According to Marshal Sergey Akhromeyev, a decision to plan a cutback in defense was made in mid-1988.[44] At this point at least, the connection among developments in the economy, the defense burden, and decisions on defense spending was clear enough.

# Chapter 6
# The Critics

Over the years CIA's estimates of Soviet military spending provoked a good deal of dissent and distrust. In this chapter we attempt to summarize and address a large accumulation of criticisms of the defense-costing effort's purpose and the methodology used to carry it out.

The critics fall into three groups. First there were the external critics. William Lee, Steven Rosefielde, and Franklyn Holzman were probably the most prominent of those who examined the agency's estimates and found them wanting. Lee and Rosefielde charged that CIA greatly underestimated the level and rate of growth of Soviet defense spending, while Holzman maintained that CIA estimates overstated Soviet outlays. A number of other academics and journalists expressed their skepticism about the usefulness and the credibility of the estimates.

Next, a series of review panels studied the defense-costing work, appraised its accuracy, and suggested improvements. As we have already noted, MEAP monitored the costing work from the panel's beginning in 1972 until its disestablishment in 1993, and a special MEAP subgroup was convened in 1982 to report on the spending estimates' usefulness to policymakers and to critique the methodology underlying the estimates. Another review panel was established in 1991 at the direction of the House Permanent Select Committee on Intelligence (HPSCI) to review CIA's research on Soviet economic and military-economic topics. These panels' reports and the extensive testimony given to the Selin and Becker panels considered nearly all of the questions that had surfaced in the 1960s and 1970s concerning CIA's defense spending estimates.[1] Finally, U.S. NATO allies also acted as critics and reviewing panels. CIA's estimates were presented at least annually at NATO meetings in Paris and Brussels. Some of these countries prepared their own estimates, and CIA analysts discussed differences at the NATO meetings or bilaterally.

To avoid repetition, the approach we take in this chapter is neither chronological nor individual. Rather, we attempt to sort the separate strands of criticism and address them one by one. We look at four broad categories of criticism: doubts about the utility of the estimates, definitional and conceptual questions, attacks on the reliability of the spending estimates, and charges that the estimates are contradicted by Soviet statistics. Some simplification of views is unavoidable, and not all of the critics are identified. Nonetheless we believe we deal with the salient issues.

## Doubts about the Estimates' Utility

The controversy about the utility of CIA's defense spending estimates boils down to these questions: Are the measures intrinsically useful? If so, do they measure what they are supposed to? Answers differ depending on whether ruble or the dollar estimates are assessed.

### The Ruble Estimates

Regarding the intrinsic utility of a ruble estimate of Soviet defense spending, there is little disagreement. As David Chu of DoD's Program Analysis and Evaluation (PA&E) shop told the Selin panel, the Pentagon's top people can't absorb all the minutiae which costs them a good deal to acquire. They need, he said, summary indicators like total defense spending and share of GNP.[2] Lee declared that what a country is willing to pay for defense is an "intrinsic intelligence question for any country."[3] He added that the U.S. intelligence community does argue about national priorities and uses the magnitude and rate of growth of defense spending in those arguments. The Working Group on Soviet Military Economic Analysis made the same point: "The ruble estimates are essential to any overall analysis of the Soviet economy."[4] Support for the view that knowledge of a country's defense budget is useful is not confined to the "blue" side of the East-West confrontation. Gen. Michael Moiseyev, chief of the Soviet general staff, said such knowledge "makes it possible to get an idea of a state's defensive capability in a condensed form which is easy for most people to understand." Moreover, using this indicator, various statistical data on a country's economic potential and mathematical models of armed forces building and development allow specialists to estimate a "series of parameters of an army's and navy's military-technical condition."[5]

Doubts about the ruble estimates center on two propositions. First, CIA's estimates in constant prices are unlikely to resemble the figures Soviet leaders would see. Thus, some contend, CIA estimates cannot be used in the analysis of Soviet decision-making. Second, Soviet prices are too deformed to provide realistic measures of the true economic burden of defense.

It is probably true that Soviet leaders never had in their policy deliberations an articulated set of defense spending figures in constant prices. To the extent that they had any figures at their disposal, they were likely in current prices, although leaders would also have access to data on the gross output—supposedly reported in constant prices—of individual branches producing military goods. These gross output figures, however, were greatly inflated and were a better representation of production in current prices than in constant prices.[6] It is likely that the Soviet statistical system, even in its highly classified compartments, never generated a full accounting of defense outlays in either current or constant prices. (Later in this chapter we discuss the basis for this statement.)

Whatever the state of knowledge about the defense budget within the Soviet leadership, the ruble estimates, if reliable, provide useful information about the place of defense in the Soviet economy. Some critics argue, however, that Soviet pricing was too chaotic to give an accurate price of the real resource cost of defense or the share of resources going to defense. A large amount of literature discusses the problem of translating Soviet prices into factor cost,[7] but the CIA in its GNP and defense estimates acted on the conviction that Soviet prices had some fundamental basis in labor costs. It found that adjustments to bring established prices closer to factor costs did not change the value of GNP or defense markedly. As we noted in chapter 5, estimates of the growth of defense spending were relatively insensitive to changes in relative prices, and the share of defense in GNP at factor cost was only slightly higher than its share in established prices. In this sense the range of uncertainty about the estimates because of distortions in Soviet pricing is not large enough to undermine the utility of the ruble estimates.

### The Dollar Estimates

Whereas the utility of reasonably accurate estimates of the ruble cost of Soviet defense spending was generally accepted, this was not the case for

CIA's estimates of the dollar-equivalent cost of Soviet military programs. Critics denounced the estimates as artificial and the U.S.-USSR comparisons of the dollar costs of military programs as poor measures of relative military power and subject to abuse.

The dollar estimates were artificial in the sense that the USSR didn't spend dollars. Moreover, the dollar-costing methodology asked what it would cost the United States to procure the Soviet forces even though U.S. military planners would not choose to build forces in the same way as the Soviets. Some critics questioned especially the application of U.S. pay rates to large military forces of the Soviet Union (and China), arguing that this practice unduly inflated the dollar value of Communist military programs; this charge was justified to the extent that differences in the quality of manpower in the USSR and United States were not taken into account. Other opponents of dollar costing were more categorical in their comments. Thus Lee, asked if his criticisms of CIA's work extended to the dollar estimates, replied, "I said earlier that as far as I'm concerned the dollar estimates are intellectually an exercise in scatology."[8] Igor Birman agreed that the dollar comparisons made little sense because their meaning was too "hazy."[9] He suggested that a system of indicators was necessary, including annual procurement costs or the value of the stock of military hardware.

Another strain of criticism also centered on the comparisons' meaning. Some wondered whether activities in two countries as different as the United States and the Soviet Union could be compared in some common denominator. Whatever its relevance to broader comparisons, the skepticism seems unwarranted with respect to defense programs. The USSR was more competitive with the United States in the military sphere than it was in other areas. At any given time the United States was procuring military hardware that the USSR had not yet produced and in many cases was still unable to manufacture, but this problem is encountered in all transnational and intertemporal comparisons. For example, to compare production or consumption in the United States in 1980 and 1996 in terms of 1980 prices requires that a 1980 price be found for goods produced in 1996 but not in 1980.

The most pervasive criticism of the dollar estimates was that they were taken as a comparison of relative military power and thus misused in the public discussion of U.S. and NATO defense policy and military budgets. Even though the estimates were—with very few lapses—presented with the warning that the cost comparisons were not a measure of relative military capabilities, these caveats tended to get lost in the debate over Soviet military goals and U.S. defense policy. For what it is worth, CIA made it plain that the dollar costs at best offered an appreciation of the general size of Soviet military programs in terms U.S. policymakers could understand. A comparison of the dollar costs of Soviet and U.S. defense programs was not in itself a measure of relative military power or force effectiveness. In every CIA dollar paper a standard disclaimer attempted to make this clear. For example, a 1973 paper stated: "Such comparisons provide rough impressions of the relative sizes of the total defense efforts of the two countries and how these efforts in broad terms have been divided among major missions and economic resource categories. The comparisons should not be used to draw conclusions concerning the relative military effectiveness or capability of U.S. and Soviet forces. Nor should such comparisons be used to draw inferences concerning relative productivities of the Soviet and U.S. economies or the internal distribution of resources."[10]

Ten years later the warning had become more elaborate:

> *Dollar costs can be used to compare the overall magnitudes and trends of the defense activities of the two countries in terms of resource inputs. Resource inputs have an important advantage over other inputs (such as the numbers and types of weapons) in that they permit aggregate comparisons. Dollar cost evaluations, for example, take into account differences in the technical characteristics of military hardware, the number and mix of weapons procured, manpower strengths, and the operating and training levels of the forces.*
>
> *But dollar valuations still measure input rather than output and should not be used as a measure of the overall effectiveness of U.S. and Soviet forces. Assessments of capability must take into account military doctrine and battle scenarios; the tactical proficiency, readiness, and morale of forces; the numbers and effectiveness of weapons; logistic factors; and other considerations. Thus, dollar valuations are useful as general indicators of changes over time in a country's emphasis on military forces. They are not sufficient to compare the overall capabilities of forces. (The order-of-battle data provided with the dollar estimates will, however, give the reader some additional insight into the relative size and composition of the two forces.)*
>
> *Dollar costs do not measure actual Soviet defense spending, the impact of defense on the economy, or the Soviet perception of defense activities. These issues are more appropriately analyzed with ruble expenditure estimates.*
>
> *It should also be noted that the Soviet dollar costs given here do not mea-*

sure manufacturing efficiencies in Soviet defense industries; they are estimates of what it would cost U.S. manufacturers to produce Soviet weapons; thus, the dollar costs for both countries are based on U.S. efficiencies.[11]

Other criticism of the dollar estimates centered on the contention that they were neither broad enough nor focused sufficiently. We noted in an earlier chapter that CIA was urged repeatedly to expand the dollar comparison to include other NATO countries and the non-Soviet WP countries. At the same time, some proposed that comparisons be focused on regions of confrontation such as Central Europe or the Soviet Far East. These excursions were attempted tentatively in the 1970s and more thoroughly in the 1980s.

The summary evaluation of the utility of the dollar estimates should perhaps be left to the principal users of these estimates—planners and officials in the DoD. In 1977 Secretary of Defense Harold Brown, in a memorandum to the DCI, declared that the reports and analysis produced on military-economics were "the basis of the comparative economic analysis employed by Defense." He added, "the dollar estimates provide the best, single aggregated measure of U.S. and Soviet defense efforts."[12] When CIA announced its intention to discontinue the dollar estimates, Andrew Marshall, Director of Net Assessment in the Office of the Secretary of Defense, wrote DDI Helgerson "to urge that CIA continue its work on dollar cost estimates of Soviet defense programs." Marshall supported his request with this statement:

> *Dollar cost data has been the major foundation for my office's work on a military investment balance report that top defense officials have found very useful in characterizing broad trends in our security situation. For all its limitations as an indicator of the military balance, dollar costing allows dissimilar systems and activities to be aggregated so that broad trends in national military capabilities can be depicted. Moreover, some activities not ordinarily compared in "bean counts" can be measured in dollars, for example hardening and sheltering programs, or command, control, and communications activities. Even intangible factors such as training and readiness can be valued at the dollar cost of the activities that foster them.*[13]

A succession of DoD representatives testified before the Selin panel in 1983 on the utility of the dollar estimates. Felix Fabian said "our clients . . . only speak dollars." Paul Berenson argued that the expenditure estimates—

at least for procurement—were measures of quantity and quality and therefore indicators of "military capability or modernization."[14] At the end of its review the Working Group on Soviet Military Economic Analysis reported that users of the dollar estimates "were unanimous in their opinion that the various players in the defense debate—the military services, the Secretary of Defense and his office, the congressional committees and their staffs, and the general public—absolutely demanded a shorthand yardstick to compare U.S. and Soviet military spending, as a surrogate for an overall comparison of capabilities."[15]

## Definitions and Concepts

Over the years three definitional or conceptual questions were raised regarding CIA's estimates of Soviet military spending. There was the long-running debate over what would be included in "defense," especially for the analysis of its economic burden. Then the index number problem led to considerable controversy over the interpretation of the agency's findings. Less noted was the conceptual difficulty caused by the choice of a resource input measure rather than an output or performance measure in costing Soviet defense programs.

### The Definition of Defense

CIA's broad definition of defense—which covered all spending on space; outlays for the KGB border guards, the militarized USSR Ministry of Internal Affairs (MVD) troops, and the railroad and construction brigades; and the costs of civil defense—was generally considered reasonable as long as it was not used in comparisons with U.S. defense program costs. Lee did quarrel, however, with CIA's explanation that the broad definition included activities that the Soviets regarded as national security-related costs. He said that "to the best of my knowledge, no one in the CIA or DIA was ever able to document the source of this definition" and indicated that a narrower definition comparable to that used in the West was preferable.[16]

Although critics had different views on the proper definition of defense, they generally agreed that different definitions should be used for different purposes. The narrower definition of defense—which included the five major military services, National Command Support, and the KGB border guards—was used in U.S.-Soviet comparisons. Henry Rowen, like Lee, thought the narrow definition was appropriate for this purpose.[17] For comparisons Birman preferred an even narrower definition that included only items "that relate to military might." He would exclude military pensions.[18] CIA's broad definition, on the other hand, attracted a good deal of criticism on the grounds that it was insufficiently broad. Marshall told the Becker panel that he would like to see some alternative definitions because one of the notable differences between the United States and the Soviet Union was the "extent to which their military establishment itself is fundamentally focused on preparing for a really big mobilization in the case of war." For this reason, Marshall suggested adding a mobilization component to the comparison. In addition he claimed that the Soviet civilian economy was influenced by military considerations to a degree unknown in the United States, citing the new trans-Siberian railroad, the Baikal-Amur (BAM), and other features of the national infrastructure as well as the mandatory maintenance of mobilization reserves.[19] Norbert Michaud, a DIA manager, worried that CIA's broad definition missed the investment in defense production facilities and the infrastructure necessary to support weapons production.[20] To the extent that military hardware prices didn't reflect interest and depreciation on invested capital, this was true, but Soviet prices were generally flawed on this account; hence the need for a factor cost adjustment.

During the 1980s the pressure for expanded definition grew. Rowen suggested adding to CIA's broad definition the costs of maintaining the Soviet position abroad, the costs of preparing for war, and the hidden subsidization of the military by the civilian sector.[21] This subsidization included the cost of standby production capacity, stockpiles of critical materials, civil defense, and the incidental costs associated with Aeroflot's wartime mission.[22] Parker and Rosefielde especially criticized CIA's "myopia" in assessing the Soviet Union's passive defense effort—civil defense, "military hardening," defense-related industrial dispersion, and the expense of maintaining manpower reserves.[23] As we have discussed, the interest in exploring broader definitions of defense led to new sets of estimates within and outside CIA.

Some doubted that a reliable definition of defense was possible in the Soviet context. They argued that the military-civil boundaries in the Soviet economy were so blurred that a clearcut definition could not be nailed down. Gen. William Odom, appearing before the Selin panel, spoke of the historical development of a military "around which all the economic activity is done and planned." He discussed the military's preemption of the best materials and other resources, which was not reflected in any account-

Critics

ing system, so that "I've said to Soviet officials, Brezhnev and Gosplan themselves—[if] they tried like fury—couldn't come up with a reasonable account, measure of, what you put in your defense. You've organized yourself to prevent it."[24] Gates, who had shared an office with Odom at the NSC, came to distrust the defense spending estimates for the same reason. As we noted earlier, he had an instinctive view that the CIA estimate of the defense burden was much too low.

In the end CIA attempts to calculate a broader version of defense did not markedly change the spending estimates or greatly alter the general appreciation of defense's impact on the economy. The materials necessary for estimating the costs of the infrastructure supporting defense are not at hand, nor are they likely to be. We should remember that a calculation along these lines should also identify and subtract the work the military did for the civilian economy—for instance, help with the harvest, railroad troops' civilian work, and the employment of construction troops in civilian factories and civil construction.

## The Index Number Problem

In any comparison of monetary values over time or between countries, a choice must be made about which price base to use. For the most part CIA used dollar prices in its U.S.-Soviet comparisons of defense programs, sparking protests from those who thought a comparison employing ruble prices should also be presented. Questions regarding index number effects were also raised by the results of the conversion of the ruble defense spending estimates from a 1970 to a 1982 price base.

RUBLES VS. DOLLARS. The dollar comparison was attacked on the grounds that simply presenting its results was one-sided. Ruble comparisons were theoretically equally valid (i.e., a comparison of the ruble estimate of Soviet military spending with an estimate of what the USSR's ruble cost would be to produce the U.S. mix of military goods and services). On standard index number theory and experience, the United States would stand taller than the Soviet Union in a ruble comparison than in a dollar comparison (see appendix B for more detailed explanation). Holzman has hammered this point home in many publications over the past twenty-five years.[25] The agency responded that it could not routinely estimate the Soviet cost of producing U.S. hardware, particularly for hardware the USSR was not capable of producing. Soviet defense enterprises could not be tasked to give an answer, although CIA could and did contract with U.S. defense industry to obtain the dollar costs of manufacturing Soviet weapons in the United States.

CIA intermittently attempted to value U.S. programs in rubles in support of a ruble comparison of Soviet and U.S. military programs. The approach taken, however, necessarily involved shortcuts and did not satisfy the agency's critics. The most recent calculation of a ruble comparison resulted in the USSR-U.S. comparisons displayed in figure 6.1. As index theory suggests, the USSR-U.S. ratio of defense programs is lower when both countries are measured in rubles than when both countries are measured in dollars. In the ruble comparison, it is true, the United States overtakes the USSR in total spending in the early 1980s rather than in the late 1980s, as it does in the dollar comparison. Nonetheless the general trends are substantially the same. Holzman in particular maintained that the difference between the ruble and dollar comparisons should be greater than

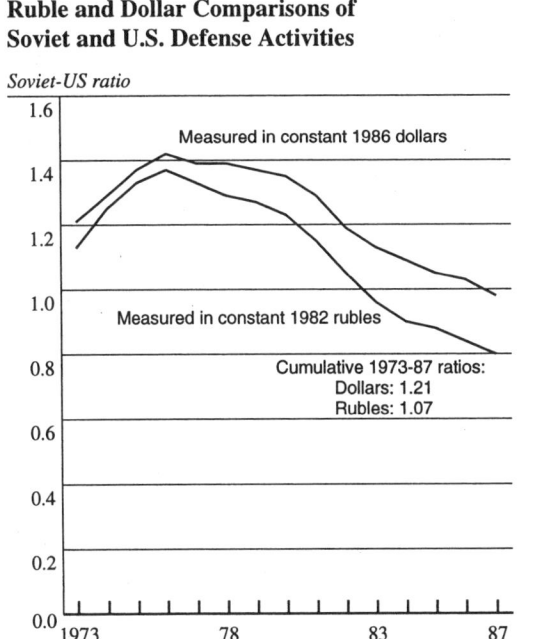

**Figure 6.1**
**Ruble and Dollar Comparisons of Soviet and U.S. Defense Activities**

that suggested in figure 6.1. He cited the differences reported by CIA for consumption and investment.

In the absence of a ruble comparison carried out with the full cooperation of Soviet authorities, we cannot reach a definitive finding on the index number relativity of comparisons of the costs of U.S. and Soviet military programs. To calculate the cost of U.S. programs in ruble prices, CIA analysts had to use ruble-to-dollar ratios derived from Soviet weapons. These ratios were then raised by 20 percent to take into account the likelihood that ruble-to-dollar ratios based on U.S. weapons—which were technologically more complex—would be higher than the Soviet-based ratios. Whether this adjustment was too little or too much cannot be determined.

There is good reason to expect that ruble and dollar comparisons would be closer in defense than in other end uses of GNP. Ruble and dollar comparisons of the volume of goods and services produced or consumed will be identical, as Holzman recognizes, if relative prices or relative quantities are the same in two time periods or two countries. Relative prices for military goods and services are obviously quite different in each country. Relative quantities of goods and services purchased by the U.S. and Soviet military are much more alike, however, than the quantities of goods and services that comprise the GNPs of the two countries. The relatively small difference between the ruble and dollar comparisons, even when more ruble prices became available, could also be due to the special nature of the prices for military goods and services. Index number relativity depends not only on differences between relative prices or relative quantities in the two countries but also on the existence of an inverse correlation between relative prices and relative quantities. This correlation may have been weak in the defense sector, with its subsidies and plant-specific pricing.[26] In addition defense decisions were driven less by relative prices than by the size and nature of opposing forces.

THE CONVERSION TO A 1982 PRICE BASE. MEAP, with increasing insistence, had urged the agency to convert its ruble estimates of Soviet defense spending from a 1970 price base to a more recent one. When the USSR conducted a major price reform in 1982 and the agency prepared to put its estimate of Soviet GNP on a 1982 base, the time for conversion of the defense spending estimates had obviously arrived. The problem was that SOVA had very few 1982 ruble prices for weapons and equipment to support such a conversion for the all-important category of procurement spending, even though a crash effort had been mounted to collect more military prices.[27] After almost four years of research SOVA found a way to deal with the dearth of prices tied to the year 1982.

Reporting to MEAP in 1986, SOVA summarized the results of the conversion in the figures shown in table 6.1. In other comparisons prepared for the MEAP briefing, SOVA noted that in 1965–84 total defense spending was 43 percent higher in 1982 rubles than in 1970 rubles; procurement was 55 percent higher in 1982 rubles than in 1970 rubles; RDT&E was 29 percent higher in 1982 rubles than in 1970 rubles; O&M was 46 percent higher in 1982 rubles than in 1970 rubles; military personnel costs were 20 percent higher in 1982 rubles than in 1970 rubles; and construction was 33 percent higher in 1982 rubles than in 1970 rubles. The implied average

TABLE 6.1

COMPARISON OF SOVIET DEFENSE OUTLAYS MEASURED IN 1970 AND 1982 RUBLE PRICES[a]

|  | *Ruble price base*[b] | *Billion rubles* | | *Percent share of total* | | *Average annual changes 1971–82* | |
| --- | --- | --- | --- | --- | --- | --- | --- |
|  |  | 1970 | 1982 | 1970 | 1982 | Volume | Prices |
| Procurement | 1970 | 26.0 | 32.1 | 47.6 | 43.2 | 1.8 | 3.7 |
|  | 1982 | 40.4 | 47.5 | 51.6 | 46.0 | 1.4 |  |
| RDT&E | 1970 | 8.9 | 15.2 | 16.3 | 20.5 | 4.6 | 2.1 |
|  | 1982 | 11.5 | 19.6 | 14.7 | 19.0 | 4.5 |  |
| O&M | 1970 | 7.7 | 12.7 | 14.1 | 17.1 | 4.3 | 3.2 |
|  | 1982 | 11.4 | 18.5 | 14.6 | 17.9 | 4.1 |  |
| Personnel | 1970 | 8.6 | 10.9 | 15.8 | 14.7 | 2.0 | 1.5 |
|  | 1982 | 10.4 | 13.1 | 13.3 | 12.7 | 2.0 |  |
| Construction | 1970 | 3.4 | 3.4 | 6.3 | 4.6 | 0 | 2.4 |
|  | 1982 | 4.6 | 4.6 | 5.6 | 4.4 | 0 |  |
| Total | 1970 | 54.7 | 74.2 |  |  | 2.6 | 3.0 |
|  | 1982 | 78.3 | 103.3 |  |  | 2.3 |  |

[a]Broad definition of outlays. Totals and percentages are based on unrounded data and may not equal the sum or percent of the rounded components shown in the table.

Critics

annual increase in prices in 1971–82 ranged from 3.7 percent per year for procurement to 1.5 percent per year for personnel. For defense spending as a whole it was 3.0 percent (compared with 2.3 percent for GNP).

Moreover, the average annual growth of procurement and total defense in constant prices was slightly less in 1982 prices than in 1970 prices (just as growth in GNP was less in 1982 prices than in 1970 prices). The defense burden (14.3 percent of GNP in 1970, in 1970 established prices; and 14.3 percent of GNP in 1982, in 1982 established prices) remained unchanged because the slower growth of real defense outlays than real GNP growth was offset by greater inflation in the defense sector than in the economy as a whole. The burden at factor cost was estimated to be about 1 percentage point higher. The estimates in the new price base were consistent with the findings revealed a few years earlier: slower growth in total defense spending after 1975 due primarily to a procurement plateau.

In explaining the methodology for estimating procurement in 1982 prices, SOVA said it had revised the earlier procedure in which only circa 1970 prices were used to estimate 1970 prices (ignoring any possible inflation that might have occurred between 1970 and, say, 1972). Instead all available post-1967 prices were used (142 in all) to make the sample as large as possible.[28] The challenge was to move these prices to a 1982 price base with the help of regression analysis. In brief, the estimated constant dollar costs of a given procurement item were taken as an anchor. Changes in the constant dollar costs of interceptor aircraft, for example, were taken to account for all quality change embodied in one interceptor compared with another. Current ruble-constant dollar ratios might vary over time because of three factors: prices for new products might be higher than those of older products in a given category by an amount not attributable to differences in quality; prices for products already in production might not be reduced in proportion to the cost reductions associated with learning; and the USSR might find it progressively more difficult (relative to U.S. experience) to produce new generations of weapons and equipment. Using multiple regression, SOVA found that the technological factor was insignificant in explaining changes in current ruble-constant dollar ratios. Instead the price depended on the time difference between the setting of the first permanent price and 1982 (inflation in the prices for new weapons) and the time difference between the year of the price quotation and the year the permanent price was established (inflation in prices for weapons in serial production).[29]

Regression analysis found substantial inflation in the prices for military hardware between 1970 and 1980. When new systems in a given category were introduced, their prices were generally higher than justified by improvements in quality, while prices for items already in production also increased substantially year after year (or failed to decline as much as past experience would indicate).[30] The results were surprising, running counter to the conventional wisdom that the Soviet military was a powerful, penurious customer. They do not seem so unusual, however, in light of the revelation that Soviet defense industrialists were the dominant party in the defense industry–Ministry of Defense relation, in decisions about what to produce and, probably, what price to charge.[31] Unfortunately the regression analysis could not take into account the impact of the 1982 wholesale price reform on prices paid for defense goods, because no post-1981 prices were included in the sample. SOVA assumed that the price reform did not affect average prices for the various defense procurement product groups and told MEAP it would continue to investigate how the price reform changed procurement prices.

As more prices for military hardware were collected, the basis for a new estimate on a 1982 price base was established. By 1990, 386 useable prices had been collected, including 101 for the years following 1982.[32] A 1990 SOVA memorandum summarized the findings of a new regression analysis of the current ruble-constant dollar ratios in the enlarged database, explaining that SOVA was more confident in its move to the new price base because total procurement in the revised 1982 prices differed from the earlier estimate by an average of only about 1 percent per year in 1970–89. It added that although total procurement did not change much, some categories of procurement did. Ship, SAM, tank, and radar prices were higher than estimated earlier, while missile, aircraft, and vehicle prices were lower. Finally, the inflation in procurement prices after 1982 was higher than SOVA had estimated.[33]

As in the original analysis underlying the conversion to 1982 prices, technology inflation (the idea that ruble-constant dollar ratios increase over time because Soviet manufacturers have increasingly greater difficulty than their U.S. counterparts producing new generations of weapons) turned out to be insignificant. The new analysis, however, found no difference between inflation in old products versus new products, so a single inflation coefficient was used to estimate the inflation rate in each procurement category.

The fact that the measured growth of defense spending in 1982 prices slowed compared with calculations in 1970 prices caused little surprise because it conformed with index number theory and experience.[34] But when the conversion of the defense and Soviet GNP estimates to 1982 prices raised the share of defense outlays in GNP, many critics expressed skepticism, if not outright disbelief. Holzman was the first to weigh in.[35] He was puzzled that the share of military spending in GNP was 15–17 percent in 1982 prices as opposed to 12–14 percent in 1970 prices. He maintained that defense hardware prices should have declined relative to those of all goods and services, and that therefore the change in the defense-to-GNP ratio was not only too large but also in the wrong direction.[36] A review panel formed on the instructions of HPSCI also could not understand why CIA's measure of the defense burden increased with the shift to 1982 prices.[37] Lee agreed, asserting that the increase was part of a cover-up in an attempt to raise CIA's estimate of Soviet defense to match his own estimates.[38]

What is at issue here is the degree of price inflation in the defense sector relative to the rest of the economy. Holzman and the HPSCI panel expected that defense, because it procures a wide range of new products whose output increases rapidly, would enjoy cost savings because of the introduction of better technology, economies of scale, and learning effects. These cost savings would be more pervasive in defense than elsewhere because defense was such a dynamic sector. When CIA changed to a 1982 price base, it found that inflation in goods and services purchased by defense was greater than in consumption, investment, and administration taken together. In particular, CIA's estimate of the overall rise in prices for military procurement between 1970 and 1982 was 98 percent, and 50 percent for the rest of defense purchases.

As we have discussed, the finding in terms of procurement depends a great deal on relating the growth in prices of successive weapons' models to the changes in their real costs, as measured by the estimated constant dollar cost of manufacturing them in the United States. At first the price sample was relatively small, and later much larger. Still, the estimating procedure was indirect, so the degree of inflation found could be subject to a fairly wide margin of error. The fact that a substantial amount of inflation was found should not surprise anyone who has followed the research on inflation in the Soviet Union. By now it has been established that inflation was especially serious in the machine-building and chemical branches of Soviet industry—the principal sources for defense procurement.[39] In the Soviet Union rapid product innovation—as in military hardware—provided an opportunity for enterprises to introduce new prices that were not justified by changes in resource cost or performance. As for the cost savings resulting from long production runs, CIA's costing procedures made extensive use of learning curves that pushed prices down as production of an item accumulated.

## *What to Measure?*

From the beginning CIA sought to measure change in the value of Soviet defense spending in constant prices, abstracting from change in the average prices of the same set of goods or resources. For most of the years in which the CIA estimates were compiled, the concept was the "constant resource price," or an estimate of what a particular good would have cost in the economic conditions obtained in a given base year. These economic conditions in turn were determined by the input prices, manufacturing technologies, and productivity levels of the base year. The constant resource price of a given good, however, could fall over time because of the effect of learning on the quantity of resources required to manufacture it.[40]

CIA's use of learning curves was questioned in the early 1980s. Michaud told the Becker panel about an "interesting problem that I want to bring up." He said that if the goal was to measure things in constant prices, learning shouldn't be taken into account; a given item should have the same price no matter when it was manufactured.[41] Rosefielde also criticized the use of learning curves in his appearance before the Becker panel, saying that it violated the standard meaning of output and biased the constant-price dollar and ruble values downward.[42] Derk Swain, replying to Rosefielde's charge, agreed that the agency was violating Rosefielde's definition of constant cost but maintained that CIA was measuring the constant cost of producing a tank in terms of inputs. In its report the working group labeled groundless the charge that CIA had used learning curves incorrectly, arguing that learning is a "real phenomena that does reduce military costs" and that CIA was applying learning curves in a "conceptually correct and careful way."[43]

Nonetheless, the use of constant resource prices to value Soviet defense was in conflict with CIA's practice for measuring Soviet GNP. The GNP measures were designed to capture changes in the volumes of output, not in the quantity of resources required to produce the output. In the late 1980s CIA calculated an alternative measure of defense in constant output prices for use in the GNP account. It found that the substitution of output prices for resource prices made little difference to the growth rates (see figure 6.2).[44] The CIA paper from which figure 6.2 is taken explained

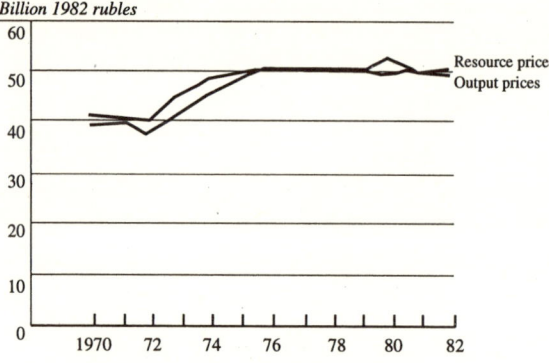

**Figure 6.2
Soviet Procurement in Two Types of Constant Rubles**

Two different measures of the annual value of Soviet procurement are shown. Both represent the levels of state set prices in the base year 1982, and both eliminate the vagaries of the Soviet price-setting system within a product group. The constant resource price line reflects cost reductions over time from learning. The constant output price line is based on the constant resource prices for the second year of production normalized to the constant resource price total in 1982. The constant resource price series is a better measure of the quantity of resources used in production, and the constant output price series is a better measure of the volume of output.

that neither measure would show consistently higher growth than the other based on a "simulation of many hypothetical procurement profiles of about the same complexity as the Soviets'."[45]

## Doubts about the Reliability of the Estimates

Attacks on the reliability of CIA's estimates can be conveniently divided into questions about the quantities underlying the estimates, criticism of the dollar costing, and doubts about the adequacy of the ruble prices for military goods and services.

### The Quantities

The basis for CIA's estimates of Soviet defense outlays was fundamentally the intelligence community's estimates of deployments and military production, supplemented by CIA estimates of operating rates, repair schedules, and construction. Although projections of production and deployments

—especially of strategic forces—aroused bitter controversy (the "bomber gap," the "missile gap," the "window of vulnerability") the estimates of past deployments, production, and the like were criticized less frequently. Thus Rowen told the Becker panel, "I think the quantities are probably pretty good."[46]

Some observers within and outside of the intelligence community noted the difficulties in establishing the physical estimates for a country as bent on secrecy as the USSR. Birman cited the "hopeless difficulties" confronting the agency in determining the quantities of arms and the quality and number of servicemen.[47] Rosefielde approached the question more analytically, asserting that CIA's accuracy in counting production varied with the size of the weapon, as the USSR tried to hide production and had probably succeeded to some extent. He judged that "for purposes of establishing an order of magnitude I believe underestimates of total Soviet procurement and production in excess of 15 percent are unlikely."[48]

Lee has been the most vociferous and persistent critic of the physical estimates—and by a wide margin. He said in 1981 that CIA didn't have accurate estimates for Soviet weapons production.[49] The next year he told the Becker panel that estimates of quantities consumed in "everyday operation were extremely difficult to estimate" and that "lots of things were going on in construction" that the intelligence community didn't know about. He added: "we are quite unable to count all of the hardware that the USSR and the Ministry of Defense procures, to say nothing of what is procured by the KGB and the MVD. We are unable, intelligence is not that good. It's that simple."[50] In a 1995 monograph Lee elaborated on his position. He claimed that CIA's estimates of weapons production were based on a "requirements" methodology that took account only of deliveries to units in the field and not additions to stockpiles. He pointed to a statement by the Soviet minister of Atomic Energy indicating that the Soviet nuclear weapons inventory was much larger than that estimated by the intelligence community, and he revisited the longstanding controversy over refire missiles.[51] His bottomline judgment was that "a limited sample of production estimates, most of which are totally unreliable and biased downward is all SCAM had at its disposal."[52]

The external panels that reviewed CIA's defense-costing work did not shed any light on the quantities underlying the spending estimates. The Working Group on Soviet Military Economic Analysis declared that in its initial stages the group had made "one very important decision"—that is, "not to review the process by which its underlying military *quantities*—

forces, manpower, items of procurement—are established."[53] Almost ten years later the HPSCI panel also decided not to review the quantity estimates, although it thought, on balance, that CIA's approach was more likely to result in an undercount than an overcount.[54]

In this book we cannot assess the reliability of the physical estimates in detail. A number of observations are warranted, however. First, over a period of almost fifty years, the intelligence community mounted a huge collection effort targeted at Soviet defense production, military deployments, and operational activity—the physical estimates. These estimates were subject to continuous review and debate. As they matured in the 1980s they were based on an accumulation of evidence. Second, while it was easier to monitor large weapon systems, these were also mainly the most expensive items, accounting for the lion's share of procurement in CIA's compilation of Soviet military spending. Third, the estimates certainly were not based principally on a "requirements" methodology. A wide range of evidence was collected and considered, and while requirements based on order-of-battle were often decisive, factors for war reserves, exports, training, RDT&E, spares, and refire (where appropriate) were taken into account. Fourth, while CIA and DIA disagree on many production estimates, the differences were mostly small; in the late 1980s the two agencies agreed on the year-by-year production estimates for more than 80 percent of more than four hundred military systems, and for the remaining 20 percent the differences were generally a question of timing, not cumulative production.

While most critics have assumed that the estimates are biased downward because of the difficulty of "seeing everything," this is not self-evident. CIA did try to fill these gaps. The items not estimated in detail, such as organizational equipment and supplies and equipage, were calculated largely on the basis of U.S. analogs. As information on Soviet logistics accumulated, it became clear that in terms of logistical tail the Soviet organization was sparer than the U.S. military. Thus the CIA estimates may be biased upward in those areas. More generally, the physical estimates of the major elements of order-of-battle and defense production were almost entirely the end result of a national estimating process in which the interests of various participants seldom induced them to minimize the threat.

Evidence of the physical estimates' reliability from Soviet sources centers on the procurement account. In the various arms control negotiations (START, CFE, and INF), the Soviet Union and the United States exchanged massive amounts of data on weapons deployments. Later the United States had the opportunity to conduct numerous inspections to verify the Soviets' information.[55] At best this evidence speaks to levels of inventory that differ from cumulative production for domestic use by operational losses, retirements, and weapons destroyed in testing. It cannot confirm or contradict the intelligence community's estimates for individual years. The information supplied by the Soviets on weapons inventories, however, is generally consistent with intelligence community estimates. The inspections have generally confirmed the information provided by the Soviets. In this sense, western estimates of the cumulative production of weapons still in the Soviet Union's inventories in the 1980s appear to be supported. At the same time, the information exchanges and the inspections do not support the idea that intelligence estimates of Soviet weapons production and procurement were systematically understated.

## Criticism of the Dollar Costing

Much of the criticism of CIA's dollar costing focuses on the procurement account, although the other dollar accounts also raised their share of questions.

DOLLAR COSTS OF PROCUREMENT. In the beginning, the extensive use of U.S. analogs tended to overstate costs because the analog was selected on the basis of a few performance characteristics without taking into account other parameters in which the analog was superior. Then the agency awarded a large number of contracts for costing individual weapons. These studies formed the basis for determining through regression analysis the dollar costs of weapons and equipment with characteristics different from those subjected to detailed costing analysis.

Critics maintained that the agency didn't know enough about the specifications of Soviet military hardware to support detailed costing; that modifications of existing weapons were wholly or incompletely accounted for; and that even when a piece of Soviet hardware could be examined closely, the performance characteristics were difficult to determine, whereas small changes in performance could be associated with relatively large changes in costs.[56] In particular they (mainly Rosefielde and Lee) believed that the costing models systematically ignored or undervalued increasing costs due to technological innovation, as one generation of weapons supplanted another or modifications were made to weapon systems.[57]

It is not possible to review the findings of or assess the methodology employed in the dozens of costing studies either contracted out or performed in-house by CIA. The dollar-costing procedures, however, were reviewed by the MEAP working group convened by DDI Gates in 1982. The principal critics of dollar costing—Rosefielde, Lee, and Holzman—presented their case to the panel. After a detailed examination of CIA's methodology, the Becker subpanel said the charge that "CIA fails to take into account technological improvement in Soviet weapons and thus underestimates the dollar prices, as well as the ruble unit values derived by transition from dollar prices, of modernized Soviet weaponry was groundless." The panel added that the "charge reflects misunderstanding of the CIA procedures, which do attempt to allow for qualitative change over time."[58]

While the panel declared the methodology of dollar costing to be sound, it said it could not judge the thoroughness with which technical change is allowed for in practice.[59] And the question simply cannot be answered. Because American manufacturers did not actually produce Soviet weapons, the estimated dollar cost cannot be compared with actual costs. But the allegation that the dollar costs did not reflect the costs of technological change can be rejected. Burton compared the cost growth of thirty-two models of U.S. tactical aircraft in 1950–80 with the estimated dollar cost growth of eighteen models of Soviet tactical aircraft during 1950–82. He found that the dollar costs of the Soviet aircraft increased at 6.3 percent per year while U.S. costs grew by 6 percent per year.[60] Indeed CIA's costing procedures resulted in very large increases in unit costs from generation to generation of weapons systems, as table 6.2 shows. With few exceptions, increasing complexity within categories of weapons resulted in steeply rising unit costs.

Marshall objected to the specifications of Soviet weapons used in the dollar costing. He thought the costs should include costs that the United States would incur if it were building a weapons system, such as the greater living space allowances on U.S. ships or air-conditioning in U.S. tanks. Marshall argued that these costs would be integral to a U.S. effort to match Soviet programs.[61] CIA maintained that this approach would violate the objective of dollar costing—to derive a price that the United States would have to pay to buy or manufacture an item or service using U.S. production practices (and efficiencies) but Soviet designs and operating practices.

CIA's costing procedures led to what was known as the "procurement paradox." David Ignatius of *The Washington Post*, for example, cited CIA/DIA testimony in which the cost of U.S. military programs was almost as much as Soviet programs' cost in 1976–80 and 1981–87, while the USSR

TABLE 6.2

COST RATIOS ACROSS SOVIET WEAPON GENERATIONS[a]

| Tanks | Ratio | Tactical antiaircraft weapons[b] | Ratio |
|---|---|---|---|
| T-55 | 1.0 | S-60 | 1.0 |
| T-55 upgrade | 1.4 | ZSU-57-2 | 3.4 |
| T-62 | 1.6 | ZSU-23-4 | 16.9 |
| T-64/72 | 4.1 | SA-9 | 3.6 |
| T-64B | 5.3 | SA-6 | 23.2 |
| T-80 | 5.9 | | |
| *Armored combat vehicles* | | *Strategic surface-to-air missiles*[c] | |
| BTR-60PB | 1.0 | SA-2 | 1.0 |
| BTR-70 | 1.7 | SA-3 | 2.5 |
| BMP | 2.8 | SA-5 | 3.6 |
| BMP-2 | 3.7 | | |
| *Artillery* | | *Attack aircraft* | |
| 122/152-mm towed | 1.0 | SU-17/22 | 1.0 |
| 122/152-mm self propelled | 5.3 | SU-25 | 2.7 |
| | | SU-27 | 3.9 |
| *Antitank missiles* | | *Fighter aircraft* | |
| AT-3 | 1.0 | MIG-17 | 1.0 |
| AT-4 | 1.8 | MIG-21 D/F | 2.1 |
| AT-5 | 1.9 | MIG-23 B | 3.8 |
| AT-6 | 3.2 | MIG-29 | 9.9 |
| AT-7 | 8.7 | | |

[a] Based on estimated Soviet costs in 1970 rubles. For each class of weapon the first model listed is the oldest. Succeeding generations are listed in descending order.

[b] Tactical SAMs include one transporter-erector-launcher and three missiles per rail.

[c] Strategic SAMs include complete launcher set and six missiles per rail. Four-rail launcher assumed for SA-3.

procured more of everything except major surface combatants. He attributed the apparent paradox to "Pentagon mismanagement, congressional meddling, the military's enthusiasm for 'gold-plated' state-of-the-art weapons that can only be purchased in small quantities, and the Soviet push during the 1970s to match U.S. force levels."[62] But most Soviet hardware was inferior to U.S. analogs, especially in user-friendliness and durability. Longer Soviet production runs also reduced the estimated dollar unit costs of procuring Soviet weapons relative to those in the United States.

DOLLAR COSTS OF PERSONNEL, O&M, CONSTRUCTION, AND RDT&E. Most of the skepticism aroused by these defense spending accounts was sparked by the underlying physical estimates. The dollar costs of Soviet personnel were a major exception. Selin, chair of the Working Group on Military Economic Analysis, observed that the weakest part of the dollar comparisons "from a public (relations) point of view if not from an economic point of view is the manpower costs."[63] Carl Jacobsen outlines the nature of the question. He noted that dollar costing is intended to show what it would cost to replicate Soviet defense programs and maintained that "the same procedures applied to Chinese forces suggests that China's military forces cost as much as those of the United States" because of the enormous size of the Chinese army.[64] The implication was that the dollar costs of Soviet personnel were too high. In contrast, Marshall told the Becker panel that the estimated dollar costs of Soviet military manpower were probably too low. He said the dollar costs of acquiring Soviet manpower were "not the same as paying actual people in uniform comparable U.S. pay and whatever immediate things go with it." Considering all of the perks that raise military manpower costs in the United States, Marshall argued the CIA methodology did not give the "best estimate of what it would cost us to have their comparative capabilities."[65] The concept of the dollar cost of military manpower was at issue. The agency insisted its concept was the only one consistent with the objective of the comparison.

Few critics challenged the O&M estimates, although they had experienced substantial change over time in procedures and results.[66] Before the late 1970s most O&M was calculated on the basis of U.S. analogs with adjustments accounting for the differences between U.S. and Soviet operating rates. Factors were applied reflecting how much maintenance was required for a given model of aircraft or class of ship. For the operating portion of O&M, the estimates of dollar costs reflected judgments about quantities consumed. The major research effort on Soviet maintenance practices (discussed in chapter 4) drastically revised the intelligence community's view of actual Soviet practice, which called for frequent maintenance of cheaply built weapons and equipment.[67] By the 1980s the accumulation of research on Soviet O&M and the collection of many Soviet operating manuals had provided a fairly solid basis for the ruble estimates of O&M. There remained the difficulty of finding the dollar equivalents for each O&M element. This was fairly simple for most of the operating portion of the account; U.S. prices for fuel, electricity, and transportation were readily available. The ruble-to-dollar ratios for major categories of weapons and equipment were also used for major spare parts. Finding a dollar equivalent for routine maintenance, however, required contract research for individual kinds of maintenance or the determination of dollar-to-ruble ratios for hours and materials devoted to maintenance, together with a calculation of U.S.-Soviet differences in the productivity of maintenance work. Dollar prices for O&M are therefore probably less reliable than ruble prices.

As was true in the O&M estimates, the differences over time in the estimates of the ruble costs of military construction and their dollar equivalents were the result of changes in the estimate of the scale of construction activity. The ruble prices for different kinds of construction were available in open sources, and the dollar prices were based on considerable research on ruble-to-dollar ratios for construction. Nonetheless, it is likely that the dollar value of a given amount of construction calculated with these dollar-to-ruble ratios would be overstated. Over the past decade, a reconsideration of the ruble-to-dollar ratio research has generally concluded that the estimated ruble-to-dollar ratios were too low because of insufficient allowances for the lower quality of Soviet goods and services. This was likely the case for the construction ruble-to-dollar ratios as well.

The estimate of the dollar equivalent of the ruble costs of Soviet military RDT&E starts with a ruble estimate that is converted to dollars. Both calculations were and are problematic. We have already discussed early and more recent approaches to the estimate of the ruble cost of military RDT&E. The transition from a methodology based on piecing together scattered Soviet reporting on research and development to one based on a ground-up approach did not change the ruble estimate much. Still, the estimate is admittedly soft.[68] This uncertainty is disturbing because RDT&E is such a large part of CIA's estimate of total Soviet defense spending and accounts for so much of its growth. One prominent student of the Soviet Union's military, for example, wondered why the ratio of military RDT&E to defense procurement was so much higher in the USSR than in the United States.[69] The answer may be that the productivity of Soviet military RDT&E was especially low.

Finally, the dollar equivalent of Soviet RDT&E posed a problem that was never really solved. It was not difficult to find ruble-to-dollar ratios to reflect average ruble and dollar pricing for R&D workers or the relative costs in rubles and dollars of the materials and investment embodied in the RDT&E process. As in arriving at a dollar value for Soviet military manpower, the assumption was that a Soviet R&D worker was just as capable as his U.S. counterpart. Given the enormous commitment of manpower and materials to Soviet RDT&E, its dollar equivalent was very large. In the 1980s CIA introduced a crude adjustment for the different productivities of the U.S. and Soviet R&D establishments by substituting the dollar-to-ruble ratio for producer durables for the weighted aggregate of the dollar-to-ruble ratios for inputs to RDT&E. The rationale for the substitution was that most RDT&E was performed in the MBMW sector and the relative efficiencies in manufacturing machinery could serve as a surrogate for relative efficiency in generating R&D "output." Soon after, the average ruble-to-dollar ratio for weapons procurement was selected as the means of converting RDT&E from rubles to dollars. In short, the dollar valuation of RDT&E shows the indeterminate character of that part of national output whose value consists of the aggregate value of inputs to an activity—whether it is education, health, or government employment (civilian or military).

### Doubts about the Ruble Prices

Skepticism concerning the reliability of the ruble prices used in CIA estimates of Soviet defense outlays centered on the prices for weapons and equipment and their spare parts. Before the mid-1970s the agency had collected relatively few useable prices for the estimates of the ruble-to-dollar ratios necessary to convert dollar estimates of procurement and the maintenance portion of O&M into rubles. When an accumulation of additional ruble prices led to a sharp across-the-board increase in ruble-to-dollar ratios for defense hardware and spare parts in 1976, the credibility of the procurement and spending estimates as a whole eroded badly. Obviously the old prices and the ruble-to-dollar ratios based on civilian analogs were out of date.

What can be said of the revised estimates in 1970 prices and the estimates in 1982 prices? First, by the early 1980s CIA had collected about one hundred ruble prices covering the major areas of procurement. In addition, ship construction costs could be derived using a Soviet costing model for merchant ships because, unlike in the United States, Soviet warships were built like civilian ships. (Ship electronics and armaments were costed separately.) The prices, however, were for different years and often of uncertain definition. They had to be converted to constant 1970 ruble prices through a statistical procedure designed to minimize the uncertainties attached to the various attributes of each individual price (complexity, the year of production, the price deflation appropriate to put it on a 1970 base). The Becker panel judged the number of prices sufficient "to generate acceptably close results" and the statistical techniques used to normalize the prices to a 1970 base a "methodologically sophisticated and intelligent approach."[70] It also vigorously supported the agency's intention to shift the Soviet defense spending estimates to a 1982 ruble price base.

When the first attempt to convert to a 1982 price base was carried out in 1985–86, more procurement prices had been collected. In all, 142 price observations during the 1968–81 time period were used to estimate ruble-dollar ratios for different categories of procurement and the average annual rates of change in these ratios. As we noted earlier, the major unknown was the effect of the 1982 reform of industrial prices on the prices for defense equipment. In the absence of any information on this point, analysts assumed that ruble prices for procurement had changed in 1982 at the average annual rate calculated for 1968–81. Then, in the late 1980s, a concentrated effort to collect more ruble prices provided the basis for a new estimate of procurement in 1982 prices (and in current prices). In the revision a total of 295 prices were used, 65 of them from the post-1981 period.

As the number of prices in the sample increased, the picture of trends in Soviet procurement changed little (figure 6.3). Replacing 1970 prices with 1982 prices, with the help of a larger collection of prices, raised the level of procurement substantially without significantly affecting the shape of the curve (part A of figure 6.3). Later the reestimate of procurement in 1982 prices, relying on a still larger collection of prices, did not affect either the level or the trend significantly (part B of figure 6.3). The distribution of procurement by mission was also essentially unchanged, even though in revised 1982 prices outlays for ships, SAMs, radars, and tanks were higher by at least 20 percent; expenditures for aircraft, missiles, and armored ve-

**Figure 6.3**
**Effect of Expansion of the Price Sample on Procurement Estimate**

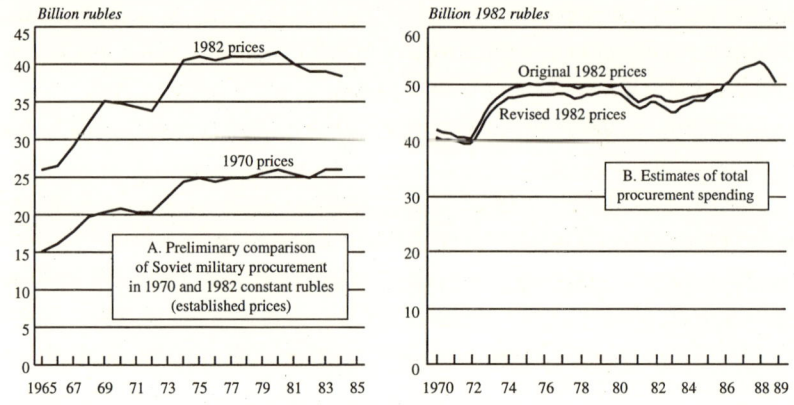

hicles declined by 8–17 percent. Successive attempts to establish a new price base suggest that, at aggregate levels, the estimates of trends in procurement at least have been robust with respect to the ruble prices supporting the calculation of the ruble-to-dollar ratios.

The 1980s research on ruble prices also shed light on a claim some critics made—that CIA's use of base-year ruble-to-dollar ratios for all years ignored a general secular upward trend in ruble-to-dollar ratios caused by the USSR's greater relative difficulty in mastering new technology.[71] In other words CIA's constant price procurement series did not capture all real growth in costs of technological innovation in Soviet weapons. (As we have shown, the estimated dollar costs of Soviet weapons did increase from one generation to another; what was not taken into account, according to the critics, was the way costs increased more rapidly in the Soviet Union.) The expanded sample of Soviet procurement prices in the late 1980s permitted a test of this proposition. Regression analysis, which included a specific technology term (the year in which a specific weapon prototype was designed), could not reject the hypothesis that increasing costs were not associated with the date of innovation. In other words the idea that the USSR found it increasingly difficult to maintain the same rate of technical progress in the defense arena as the United States was not supported by statistical analysis.

The failure to find a link between ruble procurement costs and the date of a given weapon system's design is not surprising. Throughout the

Cold War the Soviet Union trailed the West in most technologies related to defense procurement. It was always trying to close the technological gap. Its strategy, however, was usually a cautious one involving incremental advances and frequent modifications of existing weapons. When the choice of an advanced concept would have pushed up development and production costs too far, designers and defense-industrial enterprises usually found a less strenuous path.[72]

Still at issue is the charge that the ruble prices did not reflect the real cost of production. In deriving its ruble-to-dollar ratios, CIA tried to identify and use only enterprise wholesale prices, believing they were a reasonable measure of costs, even allowing for the distortions in Soviet pricing. Prices charged for arms exports and prices paid by the Ministry of Defense were excluded from the price sample on the grounds that export prices should reflect competition in the world arms market and Ministry of Defense prices were likely subsidized. One rough test of the plausibility of the ruble-to-dollar ratios derived from the ruble price sample is to compare the resulting average ruble-to-dollar ratios for defense procurement with the ratios calculated in other research for Soviet producer and consumer durables. The 1982 ruble to 1988 dollar ratios that were used in one of CIA's comparisons of defense outlays and GNP in the USSR and the United States in the late 1980s are shown in table 6.3. Whether the ruble-to-dollar ratios are aggregated using U.S. or Soviet weights, the defense ruble-to-dollar ratios are not suspiciously low.

Ruble-to-dollar ratios for defense procurement in 1982 rubles and 1982 dollars cannot be found in CIA files. An October 1986 estimate gave a Soviet-weighted 1982 ruble to 1985 dollar ratio of .67 for Soviet procurement in 1984. Like their counterparts in civilian machinery and equipment, ratios were lowest for conventional kinds of purchases—organizational equipment (.22) and engineering equipment (.28)—and highest for more technically advanced equipment—missile systems (1.25) and military space systems (1.43). The point of the comparison, however, is that the ruble-to-dollar ratios for procurement are on average substantially higher than the ratios for civilian machinery, suggesting that the ruble prices underlying the military ratios were at least not subsidized heavily. In other words, the prices used in deriving the ruble-to-dollar ratios for procurement were probably not those paid by the Ministry of Defense (and presumably reflected in the defense budgets announced in 1989 and following years). Instead they seem to be prices that enterprises actually received.

Critics

TABLE 6.3
1982 RUBLE TO 1988 DOLLAR RATIOS USED
IN A CIA COMPARISON OF U.S. AND SOVIET GNP

|                    | U.S. weighted | Soviet weighted |
|--------------------|---------------|-----------------|
| Producer durables  | .305          | .192            |
| Consumer durables  | 1.126         | .632            |
| Defense            | .476          | .439            |
| GNP                | .482          | .293            |

## Tests of Reliability

Given the numerous questions raised about CIA's estimating procedures, its assertion that the estimates had a relatively narrow margin of error—10 to 15 percent—predictably generated considerable skepticism. At times the charge was raised that the agency had failed to validate its methodology.[73] As far as the dollar estimates were concerned, empirical validation was impossible because they were analytical constructs. The ruble estimates in constant prices—whether on a 1955, 1970, or 1982 price base—were also artificial in the sense that except in the base years there was no reason to expect that accounts in the Soviet Union could validate or challenge CIA's estimates. In the absence of benchmarks to appraise the CIA estimates, they had to be judged by other criteria. For the dollar estimates of Soviet defense program costs, the agency suggested four tests: reproducibility, reasonableness, consistency, and robustness.

The test of reproducibility requires that the sources of data be explicitly documented; that the definitions, rules, and assumptions employed in the costing be set out in detail; and that the calculations be done correctly. Throughout the history of the estimates, the struggle to keep the documentation of data sources current and complete was difficult. By the late 1970s, however, the record-keeping in this regard was quite good. Definitions, rules, and assumptions governing the estimates were recorded at intervals, with detailed recitations of the sources of all quantities and prices used in the spending estimates. MEAP's recommendations were a major force behind the decision to have analysts set aside current work to undertake the laborious task of documenting the estimates. From the beginning, the oversight of the calculations was strict. A line-by-line inspection of computer listings frequently uncovered errors that were removed in successive runs. The major danger was in the reporting of the findings, as it was possible to trip over the different definitions of defense spending and make mistakes, particularly comparing the size and growth of U.S. and Soviet defense programs.

The dollar estimates satisfied a test of reasonableness if, for a given price, wage, and technology base, the dollar costs were sensitive to changes in quantities and technical characteristics; they were. The dollar estimates could also be applied over the whole spectrum of Soviet defense programs and had their counterparts in U.S. programs. Moreover, very few users of the estimates had trouble dealing with them intuitively as a reasonable expression of the trends in Soviet programs.

The estimates were consistent if, in terms of composition, they reflected a balanced (internally consistent) order-of-battle, and they did. The goal was for production estimates to be consistent with the order-of-battle, just as the estimate of number of personnel should reflect the order-of-battle. In these areas, continuing efforts to identify the tables of organization and equipment applicable to particular units led to a creative tension that improved both order-of-battle estimates and production and manning estimates. Consistency with the order-of-battle, however, is only reassuring if the order-of-battle is maintained and kept up to date. Beginning in the 1970s a major CIA/DoD initiative to automate and update the ground forces order-of-battle and the air order-of-battle drew in more resources than were applied to the costing effort itself.

Consistency in results required, for example, that "better" forces have higher costs—again, that costs reflect both quantities and quality. A more problematic test some suggested was that force decisions should appear efficient in terms of cost. Since Soviet decisions were made with a different set of relative costs in mind (to the extent that costs figured in the decisions), this was a weak test. Nonetheless, when comparable categories of U.S. data were examined, the dollar costs assigned to Soviet activities did not seem out of line, aside from the size of RDT&E relative to procurement.

Finally, CIA relied heavily on tests of robustness. The idea behind this concept is that the accumulation of evidence over the years resulting in large part (but not entirely) from the introduction of more advanced collection systems should improve the information base supporting the estimates, and thus the estimates themselves. Similarly, continuing review of definitions, assumptions, and methodologies should enhance the accuracy of the estimates. The better (more robust) the estimates are, however, the less they change with the assimilation of additional evidence.

Critics

In fact the estimates have been quite robust in the sense that as new information has become available, the estimated aggregate dollar costs have not changed much. A 1988 study discussed the range of variation in the cost estimates for the years 1970–87.[74] Table 6.4 presents CIA's estimates for the dollar cost of Soviet defense activities as they were made in each year from 1972 through 1987 (no estimate was made in 1975). The only adjustment to these historical estimates was their conversion into constant 1979 dollars for ease of comparison. (The year 1979 was chosen as the common base year because it is approximately in the middle of the base years used in the various comparisons.) Each column of table 6.4 corresponds to the estimate made in that year for the years indicated by the row labels. By reading across the columns, it is apparent that the annual updates do not dramatically change the estimated dollar costs for any given year. A comparison of the changing estimates for a given year is a rough indication of the quality of the estimates because each additional year yields new data and improved methodologies to refine the old estimates. The turnover of analysts increases the chances that any individual's blind spots or biases will in time be removed from the estimates. It can be seen that, although each year's estimate differs from the others, they are relatively stable and show no indication of consistent bias.

The bottom section of table 6.4 indexes the dollar values to the latest estimate, made in 1987. If one assumes that this estimate represents the best available information (because it reflects the largest accumulation of evidence and the most review), it provides a suitable benchmark for the preceding estimates. Earlier estimates with indexed values greater than 100 are seen as too high. Examination of the table indicates that the track record of the past sixteen years is consistent with the claim that the estimates for the total dollar value of past Soviet defense activities are subject to no more than a 10 percent error in any given year.

The relative stability in the aggregate estimates over time is partly explained by the fact that, although there may be a considerable amount of error in individual cost estimates, these do not propagate into large errors in the aggregate estimates. This is because errors in components of a total can cancel each other as the components are summed. For example, suppose that ten component estimates of roughly equal magnitude are added together to form a total, and each is subject to an individual error of $\pm$ 10 percent. If the errors are independent of each other as CIA believes, then the error in the total would be expected to be $\pm$ 3 percent.[75] As a numerical illustration of this principle, suppose each of the individual values were

## TABLE 6.4
## CIA'S HISTORICAL ESTIMATES OF THE DOLLAR VALUE OF SOVIET DEFENSE ACTIVITIES

| Estimate for year | Year Estimated | | | | | | | | | | | | | | | |
|---|---|---|---|---|---|---|---|---|---|---|---|---|---|---|---|---|
| | 1972 | 1973 | 1974 | 1975 | 1976 | 1977 | 1978 | 1979 | 1980 | 1981 | 1982 | 1983 | 1984 | 1985 | 1986 | 1987 |
| *Billion 1979 Dollars* | | | | | | | | | | | | | | | | |
| 1970 | 136 | 129 | 133 | No estimate | 139 | 130 | 122 | 127 | 125 | 129 | 132 | 132 | 127 | 136 | 130 | 128 |
| 1971 | | 129 | 134 | No estimate | 141 | 134 | 125 | 129 | 128 | 133 | 135 | 134 | 129 | 139 | 133 | 131 |
| 1972 | | | 135 | No estimate | 144 | 136 | 129 | 133 | 132 | 137 | 139 | 138 | 133 | 141 | 135 | 134 |
| 1973 | | | | No estimate | 148 | 142 | 135 | 140 | 138 | 142 | 145 | 143 | 136 | 147 | 141 | 139 |
| 1974 | | | | | 152 | 145 | 139 | 144 | 144 | 148 | 151 | 150 | 144 | 153 | 147 | 145 |
| 1975 | | | | | | 149 | 142 | 148 | 149 | 152 | 156 | 155 | 149 | 156 | 149 | 149 |
| 1976 | | | | | | | 146 | 153 | 149 | 159 | 164 | 162 | 156 | 160 | 153 | 152 |
| 1977 | | | | | | | | 156 | 155 | 161 | 166 | 163 | 156 | 161 | 154 | 154 |
| 1978 | | | | | | | | | 156 | 164 | 168 | 166 | 159 | 164 | 157 | 156 |
| 1979 | | | | | | | | | | 169 | 171 | 169 | 162 | 167 | 160 | 160 |
| 1980 | | | | | | | | | | | 175 | 172 | 165 | 171 | 163 | 163 |
| 1981 | | | | | | | | | | | | 174 | 167 | 172 | 163 | 162 |
| 1982 | | | | | | | | | | | | | 171 | 174 | 165 | 164 |
| 1983 | | | | | | | | | | | | | | 176 | 168 | 166 |
| 1984 | | | | | | | | | | | | | | | 170 | 168 |
| 1985 | | | | | | | | | | | | | | | | 173 |

# TABLE 6.4 CONTINUED

| Estimate for year | Year Estimated |||||||||||||||| |
|---|---|---|---|---|---|---|---|---|---|---|---|---|---|---|---|---|
| | 1972 | 1973 | 1974 | 1975 | 1976 | 1977 | 1978 | 1979 | 1980 | 1981 | 1982 | 1983 | 1984 | 1985 | 1986 | 1987 |
| | Indexed values (estimates made in 1987 = 100) |||||||||||||||
| 1970 | 106 | 101 | 104 | | 109 | 102 | 95 | 99 | 98 | 101 | 103 | 103 | 99 | 106 | 102 | 100 |
| 1971 | | 98 | 102 | | 107 | 102 | 95 | 98 | 98 | 101 | 103 | 102 | 98 | 106 | 101 | 100 |
| 1972 | | | 101 | | 108 | 102 | 97 | 100 | 99 | 103 | 104 | 103 | 100 | 106 | 101 | 100 |
| 1973 | | | | | 106 | 102 | 97 | 101 | 99 | 102 | 104 | 103 | 98 | 106 | 101 | 100 |
| 1974 | | | | | 105 | 100 | 96 | 99 | 99 | 102 | 104 | 103 | 99 | 105 | 101 | 100 |
| 1975 | | | | | | 100 | 96 | 100 | 100 | 102 | 105 | 104 | 100 | 105 | 100 | 100 |
| 1976 | | | | | | | 96 | 101 | 98 | 105 | 108 | 107 | 103 | 105 | 101 | 100 |
| 1977 | | | | | | | | 102 | 101 | 105 | 108 | 106 | 102 | 105 | 100 | 100 |
| 1978 | | | | | | | | | 100 | 105 | 107 | 106 | 102 | 105 | 100 | 100 |
| 1979 | | | | | | | | | | 106 | 107 | 106 | 103 | 105 | 100 | 100 |
| 1980 | | | | | | | | | | | 107 | 106 | 101 | 105 | 100 | 100 |
| 1981 | | | | | | | | | | | | 107 | 103 | 106 | 101 | 100 |
| 1982 | | | | | | | | | | | | | 104 | 106 | 100 | 100 |
| 1983 | | | | | | | | | | | | | | 106 | 101 | 100 |
| 1984 | | | | | | | | | | | | | | | 101 | 100 |
| 1985 | | | | | | | | | | | | | | | | 100 |
| Maximum | 106 | 101 | 104 | | 109 | 102 | 97 | 102 | 101 | 106 | 108 | 107 | 104 | 106 | 102 | |
| Minimum | 106 | 98 | 101 | | 105 | 100 | 95 | 98 | 98 | 101 | 103 | 102 | 98 | 105 | 100 | |
| Average | 106 | 100 | 102 | | 107 | 101 | 96 | 100 | 99 | 103 | 106 | 105 | 101 | 105 | 101 | |

Summary 1971–87

Average = 102
Minimum = 95
Maximum = 109
Standard deviation = 3.1
+/<minus> Error (95% C.I.) = 6 percent
Number of values = 131

Note: If the population mean is assumed to be 100, then the standard deviation becomes 3.7 and the +/<minus> error becomes 7 percent.

100, giving a total of 1,000. Suppose, however, the individual estimates were 91, 107, 98, 105, 102, 98, 90, 99, 97, and 108. The total would be 995, for a total error of one-half of 1 percent, even though the average error of the individual estimates was 4.9 percent. Figure 6.4 shows the reduction in overall error as the number of individual estimates increases for the case in which the individual estimates are of the same magnitude, with individual errors of ± 50 percent.

At various times CIA tried to express the uncertainty embodied in the defense spending estimates, although it recognized that without "ground truth" there was no purely statistical approach to the problem. One such attempt, reproduced in table 6.5, tried to place confidence intervals around the best estimates of ruble defense spending and its major components in 1960, 1970, and 1980. The confidence intervals in the table again illustrate the principle that aggregate error estimates expressed as percentages tend to become smaller as the number of items included in their calculation increases. Estimates of confidence intervals for total defense spending were derived by combining analysts' subjective confidence estimates for thirty-eight resource category accounts, more than twice the number included in the calculation of any of the component categories shown. When the intervals are expressed as percentages under these conditions, the interval for the total can be smaller than that of a component, as is the case for the 1960 best estimates. How justified the uncertainty statements offered by the agency were cannot be established with mathematical rigor. They were, however, not off-the-cuff judgments but the product of considerable introspection and analysis.

## CIA's Estimates and Soviet Statistics

Some critics maintained that CIA did not make sufficient use of Soviet statistics in compiling its estimates. A few argued that Soviet statistics contradicted CIA's estimates. The "statistics" at issue were official Soviet reporting on the state budget, national income, and industrial production. Proponents of relying on Soviet statistics to estimate Soviet defense spending argued that, properly interpreted, official reporting offered a reliable and inexpensive basis for estimating the ruble cost of Soviet defense programs—and one that put the estimate in prices corresponding to those that Soviet decision makers were looking at. CIA, after devoting an enormous amount of analysis to the question for more than thirty-five years, concluded that these alternative approaches to the measurement of Soviet

Critics

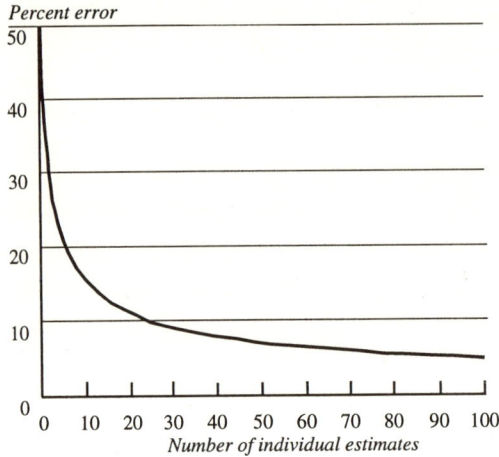

**Figure 6.4**
**Reduction in Absolute Error Using Aggregated Estimates**

Note: Each individual estimate has an uncertainty factor of ± 50 percent.

TABLE 6.5

ESTIMATES OF SOVIET DEFENSE SPENDING
AND CONFIDENCE INTERVALS FOR 1960, 1970, AND 1980[a]

|  | 1960-Best Estimate | | | 1970-Best Estimate | | | 1980-Best Estimate | | |
| --- | --- | --- | --- | --- | --- | --- | --- | --- | --- |
|  | Billion 1970 rubles | Percent of total | 90 percent confidence interval (percent) | Billion 1970 rubles | Percent of total | 90 percent confidence interval (percent) | Billion 1970 rubles | Percent of total | 90 percent confidence interval (percent) |
| Total defense spending | 26.9 | 100 | +/-14 | 48.7 | 100 | +/-9 | 70.6 | 100 | +/-12 |
| Procurement | 13.8 | 51 | +/-20 | 23.1 | 48 | +/-9 | 32.6 | 46 | +/-16 |
| Construction | 1.5 | 6 | +/-66 | 3.3 | 7 | +/-26 | 2.9 | 4 | +/-34 |
| Personnel | 5.8 | 22 | +/-16 | 7.7 | 16 | +/-8 | 8.7 | 12 | +/-11 |
| O&M | 3.3 | 12 | +/-15 | 6.9 | 14 | +/-11 | 10.1 | 14 | +/-12 |
| RDT&E | 2.5 | 9 | +/-50 | 7.6 | 16 | +/-50 | 16.3 | 23 | +/-50 |

[a]Totals and percentages are based on unrounded data and may not equal the sum or percent of the rounded components shown in the table.

defense outlays were unreliable because of Soviet statistics' incompleteness and because the USSR deliberately cooked the books to conceal the actual cost of defense programs. Even if reliable estimates could have been compiled, they would have served only as control totals and would not have provided the detail consumers wanted.

## *Defense Outlays in the Official State Budget*

Since the merger of the naval and war ministries in 1955, the Soviet Union's state budget as proposed to the Supreme Soviet and reported by the minister of finance contained a single line for "defense."[76] For many years the meaning of this budget line item inspired debate among westerners interested in Soviet military affairs. In retrospect it is hard to believe that the allocation to defense in the Soviet state budget was ever accepted as an accurate measure of the USSR's defense spending. But it was accepted in the intelligence community during the early 1950s and by other analysts over a longer time period. Whether the announced budget ever covered all the military functions and activities attributed to it in official statements, or all of a smaller but not definable portion of defense activities, remains unclear. But when in the late Gorbachev era Moscow suddenly announced that the defense budget, which had been set at 20 billion rubles in 1988, would be 75 billion rubles in 1989, and that this constituted a sum lower than that of 1988, the nature of the problem with the announced budget was evident.[77] At any rate, by the late 1950s or early 1960s almost all western students of the Soviet economy and Soviet military had recognized that the announced defense budget was not credible. For long periods of time it had remained frozen at levels that could not be reconciled with evidence of a Soviet military buildup.

For examples of reliance on the official defense budget, we can look at the estimates of the SIPRI through the 1970s and London's International Institute for Strategic Studies (IISS) in some of their annual publications on world military developments.[78] SIPRI accepted the official budget for all practical purposes while IISS said it omitted some national security-related outlays, such as some or most of military RDT&E, the KGB border guards, the military part of the MVD, and civil defense. Making allowances for these missing elements, IISS produced an estimate higher than SIPRI's and the official budget (figure 6.5).

Lee also relied heavily on the official defense budget in compiling his

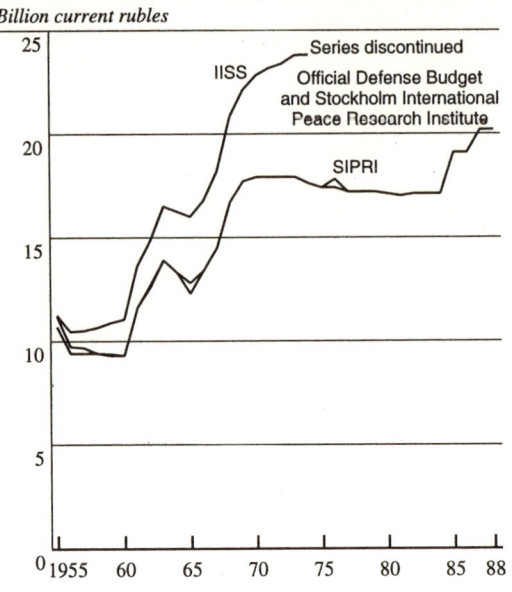

Figure 6.5
Estimates of Soviet Defense Spending
Based on the Official Defense Budget

estimates of Soviet defense spending. He assumed that the official budget covered personnel costs, O&M, and construction, but omitted procurement and RDT&E as well as the KGB border guards. Thus his estimates for the total of personnel, O&M, and construction in 1955–70 are equal to the official defense budget for most years.[79] For 1971–75 he multiplies his estimate of personnel costs by 1.48 to obtain O&M and construction combined, "the average by which the latter exceeded the former from 1968 through 1970." From 1975 on Lee simply assumes that his estimated 1975 value for personnel, O&M, and construction grows at 4.8 percent per year, the average rate of increase of the official defense budget in the 1960s.[80] His estimated growth of personnel, O&M, and construction from 1955 through 1988 is equal to the rate of growth of the official defense budget in the 1950s and 1960s.

The DIA followed another budget-related approach to arrive at an estimate of Soviet defense spending in current rubles. DIA based its estimates on the proposition that, in the ongoing budget competition, the Soviet defense establishment was limited to a fixed share of the state budget. It estimated this share in some base year and applied it to the state budget in all other years. DIA argued that the resulting series in current rubles agreed quite well with occasional statements by prominent Soviets on the level of defense spending.[81]

The state budget was a favorite hunting ground for analysts through much of the Cold War. Beginning in the 1950s, for example, CIA explored some unexplained residuals in the budget in an attempt to find defense outlays not included in the official defense budget. Some independent researchers developed budget residuals of their own.[82] This approach was pursued vigorously in the 1960s and early 1970s, and fitfully thereafter. The uncertainties were too great, and the Soviets muddied the waters by dropping out some of the detail from the budget that had aided in the research for residuals. When the Soviets announced their revised defense budget in 1989, it was relatively easy to identify the budget categories where reduced spending offset the increase in explicit defense outlays. The coverage of the new figure remained an issue, however, together with the question of whether some defense outlays were financed off-budget.

## Defense in Soviet National Income

Until the publication of the 1961 Soviet statistical handbook, almost no detail on the structure of Soviet national income was available in open sources. The tables presented in *Narodnoye khozyaystvo 1961* and subsequent yearbooks touched off a search for defense outlays in the abbreviated set of accounts reported for the first time in the postwar period. For this purpose the breakdown of national income by end use in current rubles (depicted in table 6.6) seemed promising.

Since monetary compensation to workers in the unproductive sphere is excluded from the Soviet definition of national income, researchers hoped to find a residual in "personal consumption" representing military subsistence. Several attempts were made to residualize "material outlays in scientific institutions and administration" to find values corresponding to material expenditures in O&M and RDT&E (excluding capital repair of equipment, which presumably belonged somewhere in "accumulation"). Whether military construction was included in the "increase in unproductive fixed capital" was another subject for speculation. The most concentrated attention was focused on the end-use category "increase in material

Critics

## TABLE 6.6
## PUBLISHED STRUCTURE OF SOVIET NATIONAL INCOME BY END USE

Consumption
    Personal consumption of the population
    Material outlays in institutions serving the population
    Material outlays in scientific institutions and administration

Accumulation
    Increase in fixed capital
        Increase in productive fixed capital
        Increase in unproductive fixed capital
    Increase in material working capital and reserves

working capital and reserves." Becker appears to have been the first westerner to develop the theoretical basis in Soviet writings for the "reserves" to contain additions to state reserves, which were said by some Soviets to include "means of defense of a special nature." He also pioneered in creating the residual in the increase in working capital and reserves by estimating and deducting the civilian components from the total.[83] In CIA the OER pursued the work on residuals in Soviet national income. Two papers summarized the findings.[84] A 1969 paper concluded that "the greater part of additions to state reserves might represent military outlays" and "despite the difficulties in interpretation, the behavior of additions to state reserves warrants continued study." A 1971 study attempted to residualize defense in consumption and accumulation but found that "present information is not sufficient to estimate ruble values for defense portions of national income."

By the mid-1970s much of the interest in national income residuals had evaporated. With respect to military outlays thought to be included in the consumption portion of national income, the problem was estimating a reliable residual that was a small part of a much larger total and derived only after a series of estimated deductions. The accuracy of the various deductions necessary to estimate the military part of additions to state reserves was questionable, but there was also the question of what the residual meant. According to Soviet accounting, additions to state reserves should have been net increments. If so, was weapons procurement included in the residual in gross terms or net of retirements and depreciation? Moreover, national income residuals were stated in current rubles and provided no direct measure of the real growth in the various components of defense outlays. The calculation of the residual in additions to state reserves continued to be monitored for suspiciously large changes (when the published data permitted), but not with the belief that they could be relied on. Foreign scholars joined the hunt for military outlays hidden in the national income accounts but disagreed on which outlays were included and where they could be found.[85]

## *Defense in Industrial Statistics*

Still other analysts directed their attention to Soviet industrial statistics. They hoped to isolate defense purchases of industrial products and immediately focused on the MBMW, thought to be the source of most of the military hardware procured by the Soviet defense establishment. (Almost no analysis outside the intelligence community tried to estimate the military-related production of the chemical branch of industry, although it was considerable.) Again the approach was one of residualizing—deducting machinery destined for civilian uses from the total output of machinery. The procedure depicted in figure 6.6 consisted of estimating the value of machinery output; making adjustments for net imports of machinery and deliveries to other sectors of production; and subtracting consumer durables and the machinery used in new fixed investment, capital repair, and additions to inventories.

Lee and Jeff Freeman developed this approach at CIA in the late 1950s.[86] Lee had fine-tuned his residual technique for more than thirty years, and it served as a central element of his estimate of Soviet military spending in current and constant rubles.[87] Later Rosefielde applauded the machinery residual technique in his attack on CIA estimates of Soviet defense spending, although he took a different tack in estimating Soviet military procurement.[88] Meanwhile DIA was pursuing its own investigation of machinery residuals, concentrating on the difference between the gross output of Soviet MBMW and estimates of the output of so-called civilian machine-building ministries.[89]

The difference between Lee's estimate of Soviet defense procurement, derived as a residual from MBMW output, and CIA's estimates is striking (figure 6.7). It is the basis for Lee's oft-repeated charge that CIA's estimates of Soviet military spending are contradicted by Soviet statistics. Two

**Figure 6.6**
**Residual Methodology: CIA's Nominal Estimate for 1970** [a]

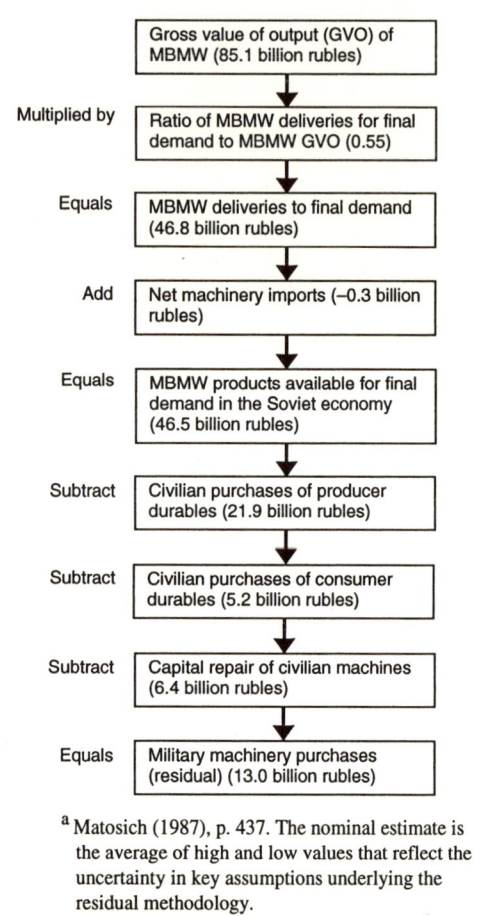

[a] Matosich (1987), p. 437. The nominal estimate is the average of high and low values that reflect the uncertainty in key assumptions underlying the residual methodology.

features of the comparison stand out. First, Lee's residual-derived procurement estimate increases far more than CIA's building-block estimate after 1965. Second, CIA's estimates in 1982 prices are substantially higher than its estimates in 1970 prices, reflecting the inflation it found in the prices of military hardware. Lee's procurement estimates in 1982 prices are lower than his estimates in 1970 prices, reflecting his acceptance of the official claim that Soviet machinery prices declined over time.

Far from ignoring the residual approach, as often charged, CIA exam-

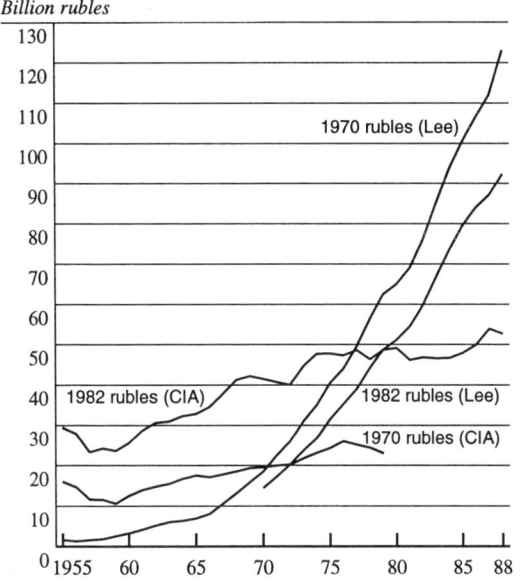

**Figure 6.7**
**Estimates of Soviet Military Procurment: CIA and Lee** [a]

[a] CIA estimates in 1982 rubles were compiled in 1985 and differ somewhat from more recent estimates shown elsewhere.

ined the machinery purchases residual repeatedly. In a 1974 paper CIA found that prospects for relying on the residual as a measure of procurement of defense hardware were "dim because of the need to depend on unofficial and ill-defined data" and because a small percentage error in such crude estimates (of the intermediate steps in the calculation) could lead to a very large percentage error in the much smaller, supposedly defense-related residual.[90] A 1979 paper, after reviewing the uncertainties involved in the residual calculation, also asserted that it was a "very rough and uncertain means of estimating the value of Soviet military hardware output."[91] The conclusion followed from the paper's calculation of three residuals, each with a range of values reflecting alternative assumptions regarding key variables in the estimate.[92] All three had a wide range of uncertainty and much lower mean values than Lee's series.

In its last machinery residual paper[93] the agency took new information into account to reestimate the residuals for 1966–84 and once again found

Critics

a wide range of uncertainty surrounding the calculations and a quite different level and rate of growth compared with Lee's estimates (figure 6.8a). Although the agency took issue with many of the assumptions and calculations underlying Lee's estimates, it reported that most of the differences in the CIA and Lee residuals could be traced to Lee's overestimates of total MBMW output (the starting point of the residual calculation), the share of MBMW output that is delivered to final demand, and net machinery imports. Correcting for these overestimates brought Lee's residuals down appreciably in level and trend (figure 6.8b).[94]

Even with the corrections, Lee's (and CIA's) residual in "comparable" prices suffers from a failure to adjust for the inflation in official Soviet statistics on machinery output. In 1974 Becker summarized the attacks on the official price indexes by Soviet academics and devised his own estimate of a machinery price index. In 1985 Kushnirsky discussed in detail the shortcomings of the official index, especially in its treatment of new products. In between, Soviets and westerners decided the official index of MBMW output was grossly inflated. Since all calculations of machinery residuals estimated in constant prices began with the official values for MBMW production, this alone would have been enough to undermine the residual methodology.

Other scholars have tried the residual approach. Michael Boretsky, in a

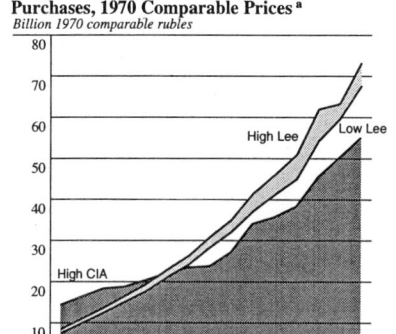

**Figure 6.8a**
Comparison of Lee's and CIA's Residual Estimates of Soviet Military Machinery Purchases, 1970 Comparable Prices [a]
*Billion 1970 comparable rubles*

[a] Comparable prices represent the Soviet method of converting industrial output from current to constant prices. These prices, however, reflect considerable inflation.

[b] Metalworking and repair were subtracted from the CIA residual to make the coverage comparable to Lee's.

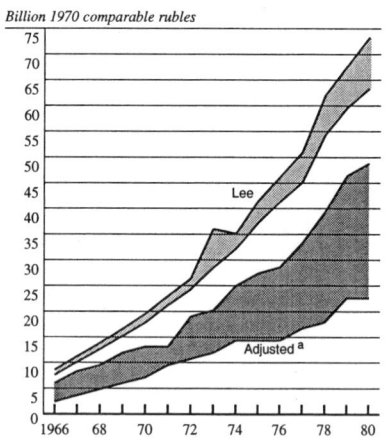

**Figure 6.8b**
Lee's Residual and Lee's Residual Adjusted by CIA, 1970 Comparable Prices
*Billion 1970 comparable rubles*

[a] CIA derived the adjusted Lee series by substituting its estimate of MBMW GVO, MBMW deliveries to final demand (excluding capital repair), and net machinery imports, leaving all of Lee's other calculations unchanged.

contribution to a 1966 JEC compendium, found a residual that grew by almost 13 percent a year between 1958 and 1963 and said the Soviet defense industrial establishment was as large or larger than that of the United States.[95] Twelve years later Stanley Cohn made his own calculations. His residuals were lower than Boretsky's and Lee's (after 1966 where a direct comparison can be made). But he noted that much of the inflation in machinery prices resulting from new product pricing would probably fall into the residual. Thus "the computed acceleration in machinery first derived in recent years may not reflect a similar trend in real military durables expenditures."[96] In 1982 Daniel Bond and Herbert Levine, building on the efforts of Lee, Boretsky, and Cohn, derived their own machinery residual for 1965–80 in current prices. It increased even faster than Lee's residual in current prices during the 1970s, but by 1980 it was little more than half of Lee's estimate.[97] According to Bond and Levine, the uncertainty surrounding the estimate was especially large after 1975 ("in the range of 50 percent or more") because of the lack of benchmark input-output data.[98] Appearing before the Becker panel, Bond said that in the absence of access to CIA's estimates, they adopted the residual approach to fill a hole in the Wharton Econometric Forecasting Associates model of the Soviet economy. But, said Bond, "we are certainly not defenders of the residual approach." In particular he didn't think the residuals could be used to check the CIA estimates of defense procurement—"too much uncertainty, too many places that you don't have data, too many price adjustments, various types of adjustments."[99]

The Working Group on Soviet Military Economic Analysis nonetheless urged the agency to continue exploring alternative methodologies to the estimation of Soviet defense spending: "In order to check overall plausibility of these (the CIA) estimates, it would be desirable to concurrently prepare a top-down gross estimate with an alternative methodology, utilizing Soviet financial and economic statistics to derive estimates of concealed military outlays in the announced reports on the state budget, net material product and output of the machinery industry. These methods have been tried in the past with anomalous results. However, it is important to continue monitoring the data sources to see whether better results can be obtained."[100] The agency tried to do this until near the Soviet Union's collapse, especially in connection with the appraisal of the revised official defense budget and plans for the conversion of defense industry to civilian production.

By the late 1980s and early 1990s a large body of evidence from Soviet and Russian sources had surfaced bearing on the various residual approaches to estimation of Soviet defense outlays. Beginning with the machinery residual, attacks on the official index of MBMW output—a key element in conducting a machinery residual—have been widespread and vehement. Without surveying the literature, we give readers an idea of the extent of the disagreement in table 6.7. It compares the official index with the CIA index of MBMW production—an independent reconstruction by Grigoriy Khanin, who is a noted domestic critic of Soviet statistics—and a revised index by a prominent Russian expert in Soviet national income and input-output statistics. The comparison shows the depth of the distrust of the official index.[101] Indeed it is hard to find in recent years a defender, aside from Lee, of the official statistics on machinery production. The recalculations reinforce a view long held by many—that attempts to derive military production from official statistics on machinery output were doomed from the start unless they took account of the inflation in these data. Moreover the Russian scholars' surrogate measures are problematic enough to prevent their use in a residualizing effort.[102]

With the increasing focus on defense conversion in the USSR and Russia, many statements have been made that bear on the size of defense industry and the share of military production in defense industry. A 1991 article by B. Rayzberg of the Gosplan Scientific Research Institute is typical. If "the overall annual volume of products of defense branches of machine building is no less than half of the overall output of the machine-building complex, which equals R 300 billion, the defense complex accounts for R 150 billion.

TABLE 6.7

ALTERNATIVE ESTIMATES OF GROWTH OF
SOVIET MACHINE-BUILDING OUTPUT
(AVERAGE ANNUAL PERCENTAGE INCREASE)

|  | 1956–60 | 1961–65 | 1966–70 | 1971–75 | 1961–88 |
|---|---|---|---|---|---|
| Official [a] | 14.2 | 12.3 | 11.8 | 11.61 | 10.0 |
| Khanin [b] | 9.6 | 8.4 | 5.7 | 5.5 |  |
| Eidel'man [c] |  |  |  |  | 5.5 |
| CIA [d] | 5.8 | 7.0 | 5.5 | 6.6 | 5.1 |

[a] Calculated from series reported in *Narkhozy*
[b] Khanin (1991), table 3.2, p. 159
[c] *Vestnik statistiki*, no. 4, 1992, p. 26
[d] GNP estimates, March 29, 1991

Eliminating the output of the consumer goods they produce (approximately 40 percent), we see that the annual military output is roughly R 90 billion per year."[103] This would be gross output of military goods in the defense-industrial ministries in so-called "comparable prices." Removing interindustry deliveries and adjusting for net exports of military hardware brings the total down to roughly 40–44 billion rubles. Adding back the defense procurement that originates in the civil ministries of machine-building and in other sectors of industry, deliveries to the Ministry of Defense fall in the range of 50–65 billion rubles. If, as seems likely, Soviet-comparable rubles resemble current rubles because of the Soviet failure to devise an accurate price index for the MBMW sector, this figure could support the CIA estimate of procurement in current rubles, but not Lee's estimate.[104]

## Glasnost and Reported Soviet Defense Spending

It has been nearly fifteen years since Gorbachev came to power in the USSR and began the changes that accelerated the decline in the Soviet state and economy. Glasnost, an increased ability and willingness to speak and write about subjects formerly treated as state secrets, was one of the forces Gorbachev set in motion, and its fruits included some interesting if not conclusive implications for Soviet defense spending estimates.

### The Revised Official Defense Budget

The official Soviet defense budget announced in 1989 demonstrated decisively that estimates relying in whole or part on the official defense budget were built on quicksand. (This applies, for example, to SIPRI's estimates and the part of the Lee estimate pertaining to personnel, O&M, and construction.) Together with subsequent budgets, the new budget provided considerable detail on the composition of the budget (table 6.8). The revised figures can be extended back in time with the help of ratios of defense to national income and the state budget that Arthur Alexander obtained on a visit to Moscow in 1991 (table 6.9). These are probably the numbers Akhromeyev was referring to in an interview for Moscow television.[105]

The new budget figures are still much lower than CIA's estimates of Soviet defense spending. But considerable uncertainty surrounds the revised budget numbers. First, there is the history of the revision itself. In late 1987, influential Soviets were telling western students of Soviet military affairs that the existing defense budget did not cover R&D and procure-

## TABLE 6.8
### THE OFFICIAL SOVIET DEFENSE BUDGETS FOR 1989 AND 1990
*(billion rubles)*

|  | 1989 | 1990 |
|---|---|---|
| Total | 77.3 | 70.9 |
| Procurement | 32.6 | 31.0 |
| Research, development, testing, and evaluation (RDT&E) | 15.3 | 13.2 |
| Personnel/operating costs | 20.2 | 19.3 |
|     monetary payments |  | 6.8 |
|     military personnel |  | 5.8 |
|     civilians |  | 1.0 |
| Material-technical supply and other |  | 12.5 |
|     food and clothing |  | 2.9 |
|     medical and communal services |  | 1.2 |
|     maintenance, operations, and repair of weapons and equipment |  | 3.3 |
|     transportation and communication |  | 1.9 |
| Construction | 4.6 | 3.7 |
|     housing |  | 1.0 |
|     sociocultural facilities |  | 0.7 |
|     capital repairs on housing and cultural facilities |  | 0.3 |
|     other (calculated residual) |  | 1.7 |
| Pensions | 2.3 | 2.4 |
| Other (production and delivery of nuclear weapons) | 2.3 | 1.3 |

ment and that efforts were under way to compile a more complete defense budget. The effort, however, would take several years; first Soviet prices would have to be rationalized to make prices of goods and services consistent across all sectors of the economy. On January 16, 1988, a Reuters dispatch from Moscow reported that Akhromeyev confirmed that the official defense budget included only salaries, costs linked to combat preparations, the use and repair of equipment, and other needs—and not procurement. He said the real budget figure could be published in two or three years when Soviet price reforms were completed. Nevertheless the new figures

TABLE 6.9

RECONSTRUCTION OF SOVIET DEFENSE BUDGETS[a]

(*billion current rubles*)

| Year | Based on national income utilized | Based on state budgets |
|---|---|---|
| 1976 | 39.8 | 39.7 |
| 1977 | 41.1 | 41.3 |
| 1978 | 43.7 | 43.7 |
| 1979 | 45.9 | 45.9 |
| 1980 | 49.0 | 48.9 |
| 1981 | 51.1 | 51.1 |
| 1982 | 53.9 | 53.2 |
| 1983 | 57.4 | 57.8 |
| 1984 | 60.9 | 60.9 |
| 1985 | 63.7 | 63.4 |
| 1986 | 66.2 | 66.3 |
| 1987 | 69.8 | 69.4 |
| 1988 | 73.1 | 72.6 |
| 1989[b] | 75.9 | 73.8 |
| 1990 | 74.6 | 74.0 |

[a] Based on figures for the defense share of national income and the state budget provided to Arthur Alexander.
[b] The published Soviet defense budget for 1989, exclusive of military pensions, was 75 billion rubles.

were released a year later before a long-discussed price reform had been carried out, suggesting that the new figures retained the weaknesses that had afflicted military pricing.

Some Soviets questioned the coverage of the revised defense budget. For example, on November 11, 1990, the Defense and State Security Committee of the Supreme Soviet met with representatives from the Council of Ministers, Ministry of Finance, and Ministry of Defense to discuss a draft defense budget for 1991 set at 66.5 billion rubles. Then on December 1, 1990, *Izvestiya*, reporting on a hearing on the 1991 budget, said that after taking into account spending covered in other budget accounts, defense

Critics

outlays would amount to 132 billion rubles.[106] Even this figure did not include spending for the KGB or MVD, according to *Komosomol'skaya pravda* reporter Sergey Bobrovskiy and a Defense and State Security Committee staffer, Igor Novoselov. Adding these additional outlays could raise the 1991 budget to 140–150 billion rubles.[107]

## Did the Soviets Know What They Spent on Defense?

There is good reason to believe that Soviet leaders did not know, and perhaps could not know, the real cost of Soviet military programs. The Ministry of Defense admittedly had difficulty compiling the revised budget.[108] In 1992, Pavel Gushvin, chair of the Russian State Committee for Statistics, reviewed the "troubled waters" of the statistics on material production he had inherited.[109] Consider industrial output in the USSR in 1990, he said. What was the cost of tanks, guns, aircraft, and other defense-related production? What was the share of the defense-industrial complex in output, employment, and fixed capital? His answer: "It is no use to search in the official statistics for a reply to this and a number of other disturbing questions. This information remained masked in the aggregate results." He added that even the chair of the Soviet State Committee for Statistics did not know the answers to these questions, and it was "as yet impossible" to find the answers "because of the methodology that was used." Meanwhile, he said, the USSR's greatest former and present politicians "on platforms or in print timidly" report an approximation of numbers like the share of defense in national income.[110]

The opacity of Soviet statistics on defense production and financing is reflected in the numerous statements made by Soviet leaders, academics, and journalists. One small group defended the new official defense budget as the real value of defense spending and denounced western skepticism as tendentious.[111] Meanwhile, many authoritative voices disagreed with the revised budget. Their statements fall roughly into the bimodal distribution illustrated in figure 6.9. Top Soviet leaders seem to have been reading from the same script—one that put Soviet military spending at or near the lower range of CIA estimates of Soviet military spending. Gorbachev told Nizhniy Tagil workers in 1990 that Soviet defense spending had reached 18 percent of national income in the 1981–85 plan period—or about 94 billion rubles.[112] Yegor Ligachev told an interviewer in 1990 that the defense budget was equal to 18–20 percent of national income—about 115–128 billion rubles if the reference year is 1989.[113] Eduard Shevardnadze, Gorbachev's foreign minister, spoke often about the Soviet defense burden. Perhaps his most definitive statement is found in his 1991 book; in it he reports telling a foreign ministry meeting in 1986 that Soviet military spending as a percentage of GNP was two and a half times more than the share in the United States—117 billion rubles if Shevardnadze was referring to GNP, or 87 billion rubles if he meant national income produced.[114] (In Soviet statistics national income differs from western-style GNP by excluding depreciation and the nonmaterial component of services—primarily the wages and profits earned in providing services. GNP was roughly one-third to two-fifths greater than national income produced during the 1980s.) Minister of Defense Dmitri Yazov, however, put the defense burden at 16 percent of GNP in conversations with Secretary of Defense Frank Carlucci—132 billion rubles if he really was referring to GNP, or 96 billion rubles if it was national income produced.[115]

Still another circle of Russian critics believes the defense burden was greater than that suggested by Gorbachev, Shevardnadze, and Ligachev. One of the earliest supporters of this view was Yuri Ryzhov, chair of a Supreme Soviet committee dealing with science, education, and culture. In 1990 he complained that the Supreme Soviet had no understanding of how the military budget was distributed among the ministries, rejecting the government's estimate. Instead of 70-odd billion, he said that "according to our experts, the figure is likely to approximate 200 billion rubles."[116] On June 7, 1990, Khanin told an *Izvestiya* interviewer that the Soviet de-

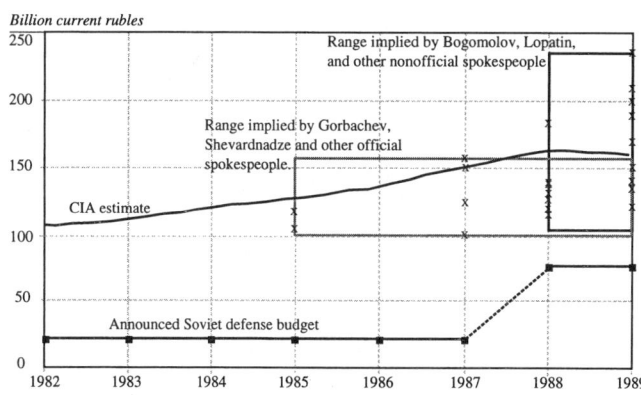

**Figure 6.9**
**Soviet Defense Expenditures**
Selected Soviet Statements and CIA's Estimate

Critics

fense complex cost 200 billion rubles a year[117]. The figure was contagious. Later that June the chief of the *Moscow News* economic section told a Japanese newspaper: "the military budget is officially set at 70 billion rubles. However, actual spending is three times that."[118] Gen. V. N. Lobov, chief of the general staff, joined the chorus, complaining, "Besides, can one call effective a defense capability which is supported by a third (and according to some estimates even more) of the gross national product?"[119]

Three observations about these conflicting Soviet declarations seem warranted. First, enough statements by top Soviet leaders exist to suggest that they believed the new official defense budget understated the real amount by one-third to two-fifths. Second, the various references to the defense burden are sufficiently vague and conflicting to indicate that even the politburo was relying on incomplete, unverified evidence.[120] Third, if the leadership itself was uncertain about the level of Soviet defense spending after ordering the Ministry of Defense to develop an honest total, little credence should be given to the back-of-the-envelope calculations of academics and journalists who had no access to the ministry's accounts, the security organs, or the defense-industrial ministries.[121]

Whether it will ever be possible to reconstruct defense spending from official Soviet statistics is questionable. Barricades include the elusive subsidies granted to defense production, the possible off-budget financing of defense programs, and the difficulty of estimating the cost of services provided free or at large discounts to the military by local governments and nondefense ministries. The subsidy issue is particularly troubling. As noted earlier, Soviet civilian and military leaders argued that a true defense budget could not be compiled until a price reform had been completed. According to Vadim Kuznetzov of the foreign ministry's U.S. desk, it would take several years (in the early 1990s) to introduce a new system of accounting in which defense industries paid their own way instead of living on state subsidies.[122] He explained, "Under the old system, a truck could cost many times less for the military than for a collective farm." Maria Shukhgal'ter suggested the extent of the distortion when she compared the share of military hardware in the final output of the Russian machine-building complex in 1990 in domestic prices (29 percent) with its share at the end of the 1980s in world prices (60 percent).[123] This kind of calculation is necessarily speculative but suggests the degree of distortion many Soviets believe was reflected in the pricing of military hardware.[124]

Off-budget financing has been raised as a possible reason for the understatement of the Soviet defense budget. Birman pioneered in this area.[125]

**Soviet Defense Spending**

Dmitri Steinberg agreed, saying that budgetary analysis convinced him that the published state budget was too small to finance civilian and military programs.[126] The defense establishment's revenue in subsidiary agricultural operations, help with the harvest, or hiring out its construction troops are examples of income that is not reported in the state budget. How extensive off-budget financing was is still a mystery. Still less is known about the services provided at no cost or at special prices by the railroads, the Ministry of Communications, and other ministries or local governments.

We have discovered that the Soviet statistical system was a patchwork of separate fiefdoms controlled by different ministries and that no integrated matrix of production, financial, and foreign trade accounts exists. If this is true, residualizing approaches to the isolation of defense outlays like Steinberg's could not work because they depend on the internal consistency of different sets of Soviet statistics. Therefore it is highly uncertain that defense spending series in nominal prices can be compiled from Soviet sources, not to mention a series in constant prices.[127]

# Chapter 7
# The Defense Spending Estimates in Perspective

We have covered most of what can be said about CIA's estimates of Soviet defense spending in earlier chapters. Here we reiterate some of our conclusions about the estimates' reliability and their contribution to the intelligence process and U.S. policy. We also consider the downside of the costing effort: its cost, the extent to which estimates were misused, mistakes and their consequences, and opportunities missed to employ the estimates in the analysis of developments in the Soviet Union during the Cold War. Finally we suggest some lessons that can be gleaned from CIA's experience sizing foreign defense programs.

## The Accuracy of the Estimates

### Salience of the Physical Description of Soviet Military Programs

In assessing the spending estimates' accuracy, we want to stress again that they were based from the mid-1950s forward on the intelligence community's estimates of the physical composition of the USSR's defense programs. As we have noted there has been some confirmation of the physical estimates, particularly in the context of the different arms control agreements with the USSR. Still, no concentrated research effort has reviewed the full array of physical estimates through all the years of compilation. Until the archives of the Ministry of Defense and the various defense-industrial ministries are made available (a distant prospect), the best description of the Soviet order-of-battle and arms production will continue to be intelligence estimates derived from a massive collection and analytic effort.

The estimates of Soviet RDT&E are an important exception to the generalization that the cost estimates relied primarily on physical evidence. Only in the 1980s was the derivation of the RDT&E estimates converted from reliance on Soviet financial statistics to a physical database. Thus, through the 1970s RDT&E spending estimates had little direct connection with a physical description of the underlying programs.

Because the evidence on order-of-battle, production, and operating rates improved and accumulated over time, the physical description of Soviet defense programs tended to become more reliable. This had two effects. First, because the estimated quantities underlying the expenditure estimates for the 1950s and early 1960s have not been systematically reviewed and updated, they are softer than those for later years. Indeed, as indicated in the reconstruction of the spending estimates, pre-1965 quantities were reexamined infrequently after the early 1970s. Second, whenever the agency reported its spending estimates, figures for the current and preceding year did not have the same foundation in physical evidence as the estimates for earlier years. A special problem in this regard was the need to take into account lead costs of weapons systems that would not enter the Soviet order-of-battle until the years after the estimate. This was a factor in the failure to make a timely call on the procurement plateau in the late 1970s.

The way evidence on physicals was collected made the cost estimates better indicators of trend than of annual changes in spending; in other words the cumulative costs of a program over a span of years were more accurate than the level estimated for any single year. For some large-ticket items—major surface ships, submarines, missile complexes—the reporting on order-of-battle was continuous, but even here the interval between procurement outlays, and deployment and the allocation of procurement costs over time for weapons whose manufacture stretched over months or years, was an issue that had to be resolved somewhat arbitrarily. The factors used to determine POL consumption and repair and maintenance schedules were even less useful for reflecting departures from Soviet planning norms, and this probably dampened the annual changes in spending estimates.

## Sizing Soviet Defense Programs

In appraising the cost estimates' accuracy, we must highlight their two major functions. The first function was to size Soviet defense programs. This was done primarily in dollars—a measure that immediately communicated an appreciation of size to U.S. policymakers. The effort was also measured in rubles so that comparisons could be made regarding spending on various defense programs in the context of the Soviet economy. The other major objective was to assess the burden or impact of defense programs on the Soviet economy and defense decision-making. The ruble estimates were used to carry out this objective.

We believe CIA did a good job overall sizing Soviet defense programs. We have presented various tests of the estimates' reasonableness and robustness, and on balance the estimates pass these tests. The fact that the comparative cumulative dollar costs of U.S. and Soviet defense programs for 1951–88 are nearly identical is consistent with the general appraisal of overall parity in size and technical quality of the two military establishments. For the ruble measures, our judgment is mixed. Before the major upward adjustments in ruble prices in the mid-1970s, the ruble spending estimates were too low. Confidence in later ruble estimates, however, is supported by the fact that the accumulation of evidence on Soviet quantities and prices barely affects the trends in estimated spending. The building-block approach's aggregation of hundreds of separate estimates meant that even though percentage errors in individual estimates could be large, percentage errors in mission or resource category subtotals were smaller, and percentage errors in estimates of total spending were smaller still. Over- and underestimates tended to offset each other. With respect to claims that Soviet ruble prices were so distorted that they could not be used for any analytic purpose, the similarity in the trends of the dollar and ruble estimates indicates that relative prices for goods and services within the Soviet defense sector were similar to those within the U.S. defense sector. They were probably reasonable reflections of underlying economic realities.

Charges that the CIA's U.S.-USSR comparisons overstated the relative size of the USSR's defense programs because the dollar comparisons were always present and the ruble comparisons were only periodically available are true in principle, but in practice this is not that important. The difference between the ruble and dollar comparisons was not large enough to alter a policymaker's sense of the trends or even the general magnitudes of effort in the two countries' military programs.

### Findings of External Review Panels

The procedures and results of CIA's defense-costing work were subjected to continuous review by MEAP from the early 1970s. Most panel members were recognized experts in Soviet studies or U.S. defense analysis. Others had high-level experience in policy agencies. Over the years they had considerable influence on CIA's military-economic research agenda. In their periodic reports to the DCI they pointed out areas for improvement, but they repeatedly endorsed dollar and ruble estimates as necessary and sufficiently reliable for use in the analysis of Soviet military trends.

As we have discussed, two additional ad hoc reviews were carried out—one at the request of DDI Gates (the Working Group on Soviet Military Economic Analysis) and the other ordered by the House Permanent Select Committee on Intelligence (the HPSCI or Millar report). After extensive hearings the working group concluded in 1983 that the costing work was competent and should be continued. The HPSCI report in November 1991 was more critical, but it was more limited in scope and less thorough than the working group report.[1] Unfortunately the working group's findings seemed to make little or no impression on agency management and received minimal distribution, while the Millar panel's report was published and briefed in unclassified forums. The agency itself was at fault, however, for not publicly replying to the Millar report.[2]

## Soviet Statements

Before the mid-1980s a few scattered assertions by Soviet spokespersons about the size of the USSR's defense budget were found in classified sources, and a few references were even made in open sources. Since then a wide range of Soviet and Russian officials, academics, and journalists have discussed the subject, generally supporting CIA estimates of Soviet defense spending in current rubles. Most of these statements contradict the official revised defense budgets that have been announced since 1989. Although some recent claims put spending at levels higher than CIA estimates, they are suspect because their authors did not have access to the classified accounts of the Ministry of Defense, the security agencies, or the defense-industrial ministries. Gorbachev, Ligachev, and Shevardnadze, whose pronouncements on defense spending tally with CIA's estimates, presumably would have had such access.

Other post-Gorbachev revelations about Soviet statistics are consistent with the CIA estimates. Soviets and Russians acknowledged that inflation in machine-building statistics was high, thus undermining estimates of Soviet defense spending such as Lee's, which relied on Soviet statistics but did not account for inflation. Statistics on the output of defense industry and the share of civilian products in this output also buttress CIA's estimates of military hardware production. Still, it is not clear that even top Soviet leaders knew the true extent of defense spending. The way budgets for national security-related institutions were compiled and the likelihood that defense activities were subsidized may have so muddled the accounting that an accurate record of defense spending from Soviet or Russian sources may never be available.

We maintain that the CIA estimates provided a generally accurate description of the dynamics of Soviet defense programs and, to a somewhat lesser degree, their levels. The estimates were rooted in intelligence collection, and they structured spending by military missions and resource categories in ways that other estimates couldn't. By their nature the dollar-equivalent levels of these program costs could not be verified, except through successive attempts to refine them and the commonsense concordance between the sizes of Soviet and U.S. military establishments and the cumulative dollar-equivalent costs involved in supporting them. The estimates of spending levels in constant prices had to contend with uncertainties about the prices being used. Neither Russian statistical agencies nor historians are likely to be willing or able to offer a better description of Soviet military spending during the Cold War until military and defense-industrial archives are opened.

## Burden Analysis

CIA's analysis of the burden of Soviet defense spending was less successful. It was plagued by the imperfections in Soviet pricing that required extensive factor cost adjustments in estimating GNP. The adjustments ideally should have been carried out for every year in current prices, not in constant prices as was the rule until near the end of CIA's costing work. To complicate matters, the degree of subsidization—in the broad sense of the word—embodied in the ruble prices CIA employed to value defense hardware was uncertain.

The goal of the factor cost adjustments was to assess the opportunity costs of defense spending—the value of civilian goods and services that could have been produced if resources devoted to defense had been directed elsewhere. In the Soviet context it is not clear that factor costs were a measure of opportunity costs. The Soviet planning and incentive systems impeded the transfer of resources and meant that their productivity differed depending on where they were employed. The agency learned a great deal about the amount and kinds of resources devoted to defense but a good bit less about their opportunity costs.

From a policy standpoint, however, the overriding justification for measuring the burden of Soviet defense programs was to understand how the burden was perceived by Soviet leaders and how these perceptions influenced leadership decisions on defense programs. Here, too, a substantial research effort fell short. Considerable progress was made in understanding the institutions involved in decisions on budgets and weapons and the time lines marking the steps in such decisions, but the impact of defense costs on these decisions remains obscure. Soviet leaders were probably looking at spending in current rubles without distinguishing between current and constant rubles, and they probably did not have a clear idea of the total cost of defense programs even in current rubles.

The intelligence community and the agency did not understand or agree on how changes in perceived costs of defense would have affected Soviet leaders' decisions. Indeed, the notion that economic constraints could limit defense spending was slow to gain acceptance within the community. Belief in the idea of economic constraints on or pressures to reduce Soviet defense budgets waxed and waned within the security policy establishment. Policymakers advocating arms control and détente tended to believe that economic pressures on the Soviet Union were working in their favor. Those suspicious of or opposed to arms control or détente generally dismissed the concept that economic pressures would deter the USSR from a spending path dictated solely by long-held military-strategic objectives.

## Contributions of the Defense Spending Estimates

It is clear in retrospect that the analytic work conducted over the years on estimating Soviet defense spending had an important impact on U.S. national security, directly through the role it played in the policy formulation process, and indirectly through the effects it had within the intelligence community on the content and quality of military intelligence analysis.

### Military Intelligence Analysis

Building-block spending analyses had their most obvious impact on estimates of Soviet forces and programs in the earliest days of their history. For example, the constraining perspective that the spending estimates provided for assessing production and deployment projections are apparent in NIE 11-6-54, *Soviet Capabilities and Probable Programs in the Guided Missile Field,* and NIE 11-5-55, *Air Defenses of the Sino-Soviet Bloc, 1955–60.* The message is clear in both NIEs that the economic implications of the forces and programs being projected cast serious doubt on the likelihood of their complete implementation.

The spending analyses' impact on the estimating process did not cease after the early days. It generally became more subtle as it was integrated at earlier stages. Godaire, for example, noted in his 1970 interview that the work of the interagency committee established to produce spending estimates for NIE 11-4-56 discovered many flaws in the available physical estimates. He cited as an example two projections submitted by U.S. Air Force intelligence: a very rapid increase in Soviet aircraft deployment and a perfectly flat projection of Soviet air forces manpower. Together the two simply made no sense to the military-economic analyst responsible for estimating spending for Soviet air forces.

A former senior CIA official told us that in the 1960s and early 1970s, when he was directly involved in drafting NIEs on Soviet military programs, he frequently reviewed historical spending patterns for insights about likely future Soviet behavior, particularly for establishing likely upper bounds for the level and pace of deployment of new weapons systems. He cited the case of expected Soviet deployment of the Moscow antiballistic (ABM) system as an example, recalling that at the time some intelligence officers from the military services were projecting ABM deployment of up to 6,000 launchers. On the basis of the estimated costs of the system and past spending patterns for other high-cost programs, he concluded that total deployment was unlikely to exceed 600 launchers even if the Soviet military was satisfied with the system's technical capabilities.[3] The most likely deployment levels presented in the pertinent NIEs of the period were much closer to the 600 level.[4]

This kind of integrative analysis took place not only at the interagency level in the context of producing NIEs, but also at the working analyst level within CIA. The interactions took place at many levels and in many forms. One of us, Noel E. Firth, who began his career in intelligence as a military-economic analyst, recalls that consultations with counterpart military analysts to resolve anomalies, inconsistencies, and gaps—apparent only because the spending analysis was being done—were daily events. Military analysts did not always welcome a second analytical look at a given problem taken from the costing perspective, nor did military-economic analysts especially welcome flaws in costing that were inevitably pointed out by the military analysts. But the interactive process that the building-block method required unquestionably improved the rigor and quality of the overall U.S. military intelligence effort. One consequence was the focus of attention on the day-to-day operating aspects of the Soviet military establishment. This attention paid off not only in more reliable estimates of operating costs, but also in an increased understanding of training levels and other readiness factors that impinged on the Soviet forces' overall capabilities.

The relation between spending estimates and force analyses most widely discussed among intelligence analysts was the tired-arm effect. In this phenomenon, observed in the early days of building-block analysis, the slope of the spending trend line declined as estimates of procurement for any class of military hardware—aircraft, missiles, tanks—extended into the future. Military analysts could not identify the series production of new systems more than a few years into the future. This created a dilemma when the objective was to project spending levels. Given past behavior it would be judged likely that established spending trends would continue; new systems, however, that would support a continuation of past trends had not been identified. Many military, military-economic, and scientific-technical analysts grappled with this problem over the years.[5] By the end of the 1970s the tired-arm effect had given way to a ramp effect, in which analysts' projections outstripped reality. By introducing probability assumptions into the projections and using Monte Carlo simulations, more reasonable spending aggregate projections were formulated.

The insights on Soviet military engineering and production capabilities provided as byproducts of the dollar-cost analyses of Soviet weapons conducted by U.S. weapons manufacturers were an unexpected benefit of CIA's military-economic effort. As the following statement by Marshall shows, the broader applications for understanding Soviet military capabilities these technical insights conveyed was recognized and appreciated in the Office of the Secretary of Defense: "I was very impressed by the Hughes SA-3 Systems Cost Study. While I recognize the inherent problems of costing out Soviet systems, I thought the paper was one of the most interesting pieces of analysis I have read in many months. Of particular value were the insights . . . into the technical capability of the Soviet designers, the comparison of their technical capabilities to ours, and the reasons why certain design options may have been chosen. Much of the analyses written on

Soviet systems is hampered by the lack of such insights, which can clarify popularly held beliefs about the Soviets and put our evaluations of their weapons into perspective compared to our own."[6]

Perhaps the least understood contribution of the military spending effort was the rigorous analytic framework it provided for organizing and preserving the essential elements of CIA's military analysis. As the building-block system developed, conceptual definitions and costing methodologies were refined, articulated, and recorded as part of the system's documentation to provide continuity from one generation of analysts to the next and to minimize ambiguity. The system provided a home for historical, current, and projection databases on the key physical parameters of the Soviet military establishment—that is, order-of-battle, manpower, and production—in addition to the spending estimates. The definitional clarity and internal consistency of this information made it invaluable for a broad range of military analyses. It was particularly helpful, for example, for military intelligence problems related to arms control where conceptual and definitional precision and data consistency over time were of paramount importance.

## The Policy Arena

The national security policy formulation process can be divided into the two intertwined, but functionally distinct, activities of analysis and advocacy. CIA's estimates of Soviet defense spending, like all products of the intelligence community, could be (and were) used effectively to support both activities.

The flow of detailed estimates of Soviet military spending from the agency to Pentagon analysts, which began as a trickle in the 1950s and reached flood levels by the mid-1960s, provides one measure of the spending analysis contribution to the policy process. In some instances the CIA contribution initially served as an input to further internal Pentagon analysis and only ultimately was brought to bear in the political arena, either explicitly or implicitly in support of a particular policy option. In other instances CIA's Soviet military spending information was used directly to support advocacy of a particular policy line.

Kaplan gives an excellent example of how the intelligence on Soviet military costs was used initially as an input by Pentagon analysts and then ultimately as ammunition in a policy debate. He shows that an internal Pentagon interactive cost-effectiveness analysis of Soviet and U.S. strategic forces, using estimated Soviet cost data provided by CIA military-economic analysts, convinced Secretary of Defense McNamara that U.S. pursuit of a damage-limiting strategy was futile.[7] McNamara then used the Pentagon analysis to argue against the large ABM and open-ended offensive weapons programs such a strategy implied. Decisions on these programs involved billions of dollars. The appearance of CIA estimates of Soviet military spending in Secretary of Defense Posture Statements and the DoD publication *Soviet Military Power* provide good examples of direct use for advocacy purposes. These documents emphasized elements of spending that indicated a greater or more rapidly growing Soviet effort.

Not surprisingly, the Pentagon placed the greatest demands on the agency for military-economic intelligence on the USSR, but it certainly was not alone. Indeed, all institutional players in the national security process with an interest in the U.S. defense budget, both in Congress and the administration, were users of or affected by intelligence on Soviet military spending. In a nationally televised speech in November 1982, for example, President Reagan cited CIA estimates of Soviet defense spending as part of the rationale for his proposed defense program.[8] President Carter made similar use of the CIA estimates in 1979.[9]

Against this backdrop the proposition can be advanced that the CIA numbers on Soviet defense spending were among the more important estimates produced by the agency during the Cold War. This is not a provable proposition, but some pertinent observations and speculations may assist in putting it in perspective. Marshall, longtime influential Director of Net Assessments in the Office of the Secretary of Defense, noted in a September 1976 memorandum: "The Secretary's Posture Statement and Defense Report for FY 1978 are to be presented in mid-January. Some of the trends and part of the argument will rest heavily on the economic analyses of the Soviet defense program performed by the CIA. . . . these economic estimates are among the best aggregate measures of relevant trends in the Soviet military efforts available. Most of the criticisms seem to misunderstand the logic of the dollar estimates. They are very useful and perfectly sound as a measure of the comparative size of U.S. and Soviet efforts."[10]

Another pertinent observation arises from the controversies that surrounded the estimates for much of their history. It is difficult to answer the question, "Why all the fuss about CIA's estimates of Soviet defense spending?" without conceding that they must have had a substantial impact on U.S. policy. It is instructive to speculate on what might have been the prevailing image of levels and trends in Soviet military activities if the CIA estimates had never been done. Perhaps the consensus view of U.S. policymakers would have been more benign and the United States would have spent less on defense. But this scenario is unlikely. Given the abundant physical evidence of the robustness of the Soviet military establishment and the rhetoric this evidence stimulated, the opposite probably would have occurred. We believe the odds are high that the prevailing view of Soviet military programs would have been more alarmist and U.S. defense spending during the Cold War would have been much higher.

This judgment should not imply that CIA's military-economic analysis in some way systematically failed to reflect the physical evidence of Soviet defense activities. On the contrary, because the work reflected available evidence about all activities in all of the forces, it provided an image-balancing and moderating function that probably would have been missing otherwise. Besides costs, no measure was available to effectively take into account all the disparate goods and services—in both their quantitative and qualitative dimensions—that comprise a massive, diverse activity, such as the Soviet military establishment. Key is the point that, given CIA's building-block methodology, their spending estimates were driven almost exclusively by the intelligence community's understanding of the underlying Soviet military forces and programs.

### Contribution to Historical Analysis

CIA's defense-costing analysis was intended as an intelligence contribution to U.S. policy deliberations during the Cold War. In the future it will help historians understand developments in the postwar Soviet Union. Together with CIA's Soviet GNP estimates, they provide a continuous and internally consistent statistical record for research on Soviet history after World War II. Given the current demands on Russian statistical agencies and the secrecy still surrounding defense and military-economic archives of the Soviet Union, CIA's estimates may well be the economic records of choice for historians.

## The Downside

We have set out what we believe were the benefits of CIA's defense-costing work, but the costs should be considered too. Isolating these costs is not easy because what is at issue are the incremental costs of doing the estimates. The military-forces analysts who would have been studying order-of-battle and production in any event cannot be counted.[11] A rough estimate of the incremental direct costs from 1955 to 1990 based on our review of historic contract and manning level files would be $25 million: $20 million in salaries for about seven hundred years of analyst time and $5 million in external contracts. Adding the overhead costs that would be incurred in support of the analysts and administering the contracts would perhaps bring total costs to $35–40 million—a modest price tag for a thirty-five-year analytic effort.

While the expense of costing Soviet defense programs was of concern to some budgeteers, the critics, who in the main accepted the estimates as sufficiently reliable on their own terms, often argued that they were misused in a manner that partly or wholly vitiated their usefulness. The so-called politicization of the estimates was supposedly one example of misuse. Inevitably the estimates tended to be embraced or rejected depending on whose agenda they helped at a given moment. But these are the facts of life for any analysis that might have an impact on policy. The real question is whether the construction of the estimates or their transmittal to the policymaking community was in any degree politicized. We found no evidence of distortion in the construction of the estimates and, with troubling exceptions regarding some of the more subtle forms of politicization exercised in the 1980s—such as senior management's reluctance to pass the estimates along when the message became inconvenient to the administration—we do not believe politicization was a serious problem.

More often the estimates were said to be misused in the sense that the dollar costs were misrepresented as indicators of Soviet defense capability vis-à-vis that of the United States. After reading all classified and unclassified papers on the dollar costs of Soviet defense programs and CIA's testimony before the JEC, we have not found any basis to this claim.[12] Over time the agency's disclaimers on this score became so lengthy and vehement, they obscured the essential point that annual outlays on defense are an extremely important input to defense capability. We also have not uncovered any instances in policy debates where the estimated dollar costs of the Soviet defense program were taken as anything more than an element of the evolving strategic competition between the USSR and the United States.

It is true that the agency allowed itself, because of the demand for current reporting and the requests of congressional committees, to be put in a position where undue emphasis was placed on estimated changes in Soviet defense spending in the current or immediate past year. These were the weakest parts of the estimates. Consumers would have been better served by more emphasis on the trends leading up to the current year—for example, average annual growth over the past five years.

The real mistreatment of the estimates stems from the few occasions when the agency rejected them—when they were bureaucratically or politically inconvenient. The delay in announcing the findings on a slowdown in the real growth of Soviet defense spending beginning in the mid-1970s, the attempts to ditch the dollar comparisons in the 1980s, and the failure—particularly in the later years—to take military-economic analysis into account in the force projections in the NIEs for strategic and general purpose forces were missed opportunities to contribute to policy decisions.

## Some Lessons Learned

The experience estimating Soviet defense outlays offers some lessons for intelligence analysis. First, an analytical path has been set up in the event that it again becomes necessary to size the defense programs of major powers. The framework is in place, and an array of methodologies have been developed. Less elaborate estimation models are available if a full-scale building-block type approach is not appropriate. In a world where some defense budgets are misleading, it does not take much imagination to envisage a resumption of defense-costing work by some U.S. government entity.

Over the often stormy years the agency learned—if it needed any additional instruction—that in the defense-economic arena, keeping intelligence judgments from being skewed by consumers' policy interests requires continuing attention. Much of the criticism directed at the estimates was inevitable because of the role they played in the debate over the U.S. defense budget. Those believing the USSR was striving inexorably for military superiority thought the agency was greatly understating the level and rate of growth of Soviet defense outlays. Others subscribing to a more benign view of Soviet intentions believed the agency's estimates exaggerated the scale and pace of the USSR's military programs. It is not hard to find fault with parts of the CIA analysis, but to abandon the estimates because they attracted criticism would be a know-nothing stance suppressing much of intelligence analysis.

The spending estimates' history teaches us that analysts must always be ready to take new information on board and to entertain unconventional ideas of the world in a timely way. The abrupt increase in the ruble estimate of Soviet defense spending in the mid-1970s, the delay in recognizing the break in the trend of these outlays after 1975, and the failures of military force projections illustrate this important point.

The history also underscores a longstanding problem in intelligence—the difficulty of integrating political, military, technical, and economic analysis despite organizational initiatives to break down barriers across disciplines. Too often the analyses flowed within their separate channels, undisturbed by research and insights running in parallel streams. The blending of different perspectives on developments in the Soviet Union tended to be joined together at the late stages of an intelligence paper or estimate, without sufficient attention to how judgments proceeding from one perspective should be modified by other points of view.[13]

In commenting on a draft of this section, Bruce C. Clarke, former deputy director for Intelligence and former director of Strategic Research of CIA, made the following observations, with which we are in total agreement:

*It seems to me that lessons to be learned from the agency's 30 years' assault on the problem of estimating Soviet defense spending include the following—*

*—War as the continuation of diplomacy by other means is a real and continuing factor in the formulation of U.S. security policy; it did not end with the implosion of the Soviet empire. Russia may be a humbled and crippled giant at this point, but it is a giant armed with weapons of mass destruction all the same and in time it will recover its strength, its sense if national mission, and its insistence on international recognition. And China today is also a giant, and a growing one, armed with weapons of mass destruction. . . .*

*This means that the existing and future force levels of at least these two countries will be a matter of enduring concern to the making of U.S. foreign policy and the determination of U.S. military policy and force levels. And this in turn means that U.S. policymakers will be asking essentially the same kinds of questions about Russian and Chinese military spending and their relation to the Russian and Chinese force level development that was true for Soviet defense spending. And they will want essentially the same kinds of answers, only better.*

*—Second point: The history of the agency's program to analyze Soviet spending amply demonstrates how long and tedious this intellectual undertaking can be, conceptually, evidentially, computationally, etc. Years are required to be ready to give good answers when they are needed. When the nukes start*

> to fly, the tanks start to roll, and the landing craft are launched, it's too late to begin creating the necessary data bases and methodologies. But this is true of U.S. force level determination too. Long lead times are involved and they must proceed with at least some rationally derived view of what they may be up against ten years hence.
>
> —Third and last: Our concern here is with what the President and his advisors need effectively to formulate the foreign and military policies that assure the well-being of the Republic in the era of weapons of mass destruction. Because of the profound budgetary implications involved, the Congress has an equal claim for substantive, informed military intelligence analysis and judgement. And the agency's experience throughout the decades since 1947 fully demonstrates that military intelligence analysis at the national level, where the Director of Central Intelligence is critically responsible for the needs of the President and the Congress, is too important to be left to the military.
>
> Time and time again, as I made my rounds to the White House, to the Office of the Secretary of Defense, and to the Hill as Director of Strategic Research, it was repeatedly impressed on me that, next to the agency's demonstrated professional and substantive expertise in military analysis, the most important quality we possessed was our organizational freedom from departmental budgetary concerns. None of our users was worried that we were likely to be skewing the analysis in favor of this or that U.S. weapons system. To a substantive question, they could expect a substantive answer, driven by evidence (or lack of it) and analysis but free of departmental spin.[14]

This last point about who performs military and military-economic analysis is crucial. As one of the authors noted in a letter to the editor of the *Washington Post*,

> I am convinced—having spent more than thirty years closely observing the dynamics of CIA relationship with the military in producing foreign military assessments—that the agency's participation in the process has saved the U.S. taxpayer many billions of dollars, contributed significantly to maintaining reasonable stability in the world balance of nuclear forces, and made nuclear arms control agreements possible.
>
> The point is not that the CIA analysts are universally smarter or better than the analysts of the military services and the Defense Intelligence Agency. Clearly, neither group has a monopoly on analytic skills. The point is rather that military intelligence analysis conducted by military organizations is

*inevitably driven by the fundamental imperative of the military commanders whom they serve. This imperative is to ensure the U.S. capability to achieve victory in the event of a military conflict.*

*This creates a tremendous incentive throughout the DoD establishment to maximize ("worst case") the military threat of potential adversaries in order to justify sufficient superiority to ensure victory. But, it is a clear case when more than enough to do the job is not necessarily better. Indeed, at the national policy level the DoD approach to military intelligence analysis left unchecked by the competitive civilian analysis . . . invites economic and perhaps even military disaster.*

*The forty year history of the cold war is replete with examples of overstatement of the threat by DoD intelligence organizations. . . . Had the U.S. reacted fully to these . . . overstated threats . . . the fragile stability of the East-West military balance could have been upset and the cold war could have ceased to remain cold. At a minimum, U.S. defense spending would have been substantially larger than it was.*[15]

# Appendix A
## Costing Improvements, 1975–90

This appendix describes the most important improvements introduced in CIA's military-economic analysis during 1975–90. It includes information on new evidence and new analytic techniques, organized by the major resource categories of spending.

### Personnel Expenditures

The military manpower estimates were the foundation of the estimates of Soviet military personnel costs. A CIA/DIA study in 1975 furnished the first thorough reassessment of Soviet military manpower since the late 1960s. OSR, after two years of research, published a report that put military manpower at 4.64 million in 1967 and estimated that it had increased by 490 thousand since 1968.[1] Some of the softer manpower estimates were those for the construction and railroad troops and the internal security troops of the MVD. These areas were targeted in three 1980 studies.[2] They concluded that the military construction units had 600,000 military and 25,000–30,000 civilians assigned them, while the estimate for railroad troops was raised from 75,000 to 130,000. The estimate for MVD internal security troops also increased from 174,000 to 193,000. By 1981 the estimate of total armed forces personnel (military and civilian) in 1971 had been raised to 5.92 million (and to 6.06 million in 1980).

Until the 1980s military manpower was calculated mainly by multiplying the number of identified and inferred units by manning factors. There was a great deal of discussion about how many of the divisions were in various stages of readiness, but if the question of an overall manpower constraint arose it was dismissed as not binding. Adverse demographic trends in the 1980s, however, altered the picture. The number of males reaching draft age was declining, and an increasing proportion of these were in Central Asia or the Caucasus, where draftees often did not speak Russian and had less education. At first OSR doubted that this demographic dilemma would lead to a reduction in military manpower.[3] But information on deferment rates for education and health reasons coupled with the demographic trends led some SOVA analysts to question whether military units could continue to be manned at previous levels. The debate within SOVA simmered until October 1987, when analysts were assigned to a special task force to reassess military manpower—with the goal of reconciling existing estimates with the availability of draft-age males. In February 1988 Dennis Nagy, DIA's Assistant Deputy Director for Research, suggested to Douglas MacEachin, OSR's director, that an interagency working group on manpower be established similar to the one in the 1970s. MacEachin accepted, but the working group never materialized. Instead SOVA conducted the review unilaterally and the results appeared in a 1990 paper.[4]

Citing new information, the new estimate for 1965–89 cut its estimate of military manpower in the 1980s by almost one-fifth. It posited that manpower had probably peaked at less than 5.1 million in 1988, compared with the previous estimate of more than 6.1 million for that year. Since the estimates of the number of units had not changed appreciably, the new manpower estimates implied a pronounced hollowing out of Soviet military forces in the 1980s, in response to demographic pressures and the leadership's disinclination to build down the force structure. The new estimates' portrayal of the composition of the manpower reductions is depicted in table A.1. Figure A.1 compares the trends in the old and new estimates.

In calculating the compensation of Soviet defense manpower, CIA could rely on a large accumulation of basic records dating to before the 1950s. The information was reviewed again in 1981 in a research paper that reported high confidence "in our knowledge of compensation for conscripts and junior officers through the rank of major," but somewhat less confidence in estimates of compensation for senior officers and those in rear services positions. Estimates for earlier years were considered more reliable because a change in compensation rates could go undetected for years.[5] In addition estimates for pay had historically been better founded than those for food and clothing. Thus a new estimate of military clothing costs reduced the estimate by 40 percent in 1979, while a revised estimate of ration costs

Appendix A

### TABLE A.1
### ESTIMATES OF SOVIET MILITARY MANPOWER FOR MID-1988[a]

| Force | Old estimate (thousands) | New estimate (thousands) | Percent change |
|---|---|---|---|
| Ground forces | 2,035 | 1,610 | -21 |
| Air forces | 610 | 480 | -22 |
| Air defense forces | 575 | 545 | -5 |
| Strategic rocket forces | 315 | 315 | 0 |
| Naval forces | 435 | 435 | 0 |
| Command and support | 1,475 | 1,110 | -25 |
|     Construction troops | 805 | 645 | -20 |
|     Railroad troops | 235 | 150 | -36 |
|     Civil defense troops | 50 | 40 | -14 |
|     Other | 385 | 275 | -29 |
|     KGB border guards | 215 | 225 | 5 |
|     MVD internal troops | 460 | 325 | -29 |
| Total uniformed personnel | 6,125 | 5,045 | -18 |
| Less KGB, MVD, and railroad troops[b] | -910 | -700 | |
| Total armed forces | 5,215 | 4,350 | -17 |

[a]Totals, subtotals, and percentages are based on unrounded data and may not equal the sum or percent of the rounded components shown in the table.

[b]The Soviets do not consider these militarized troops as part of their armed forces. CIA counted them in its estimate of total military manpower, however, because they consist primarily of uniformed, conscripted personnel.

raised the old estimate by 10 percent in 1980.[6] Like most of the defense spending estimates, the per-man compensation rates employed by CIA probably will not be contradicted or confirmed soon. One small piece of evidence came from records of a Soviet tank division in the 1980s. These records showed a per-man cost of 1,400 rubles per year (pay, food, travel, medical care, and the like) compared with SOVA's independent estimates of 1,570 rubles for a similar division.[7]

## Costing Improvements

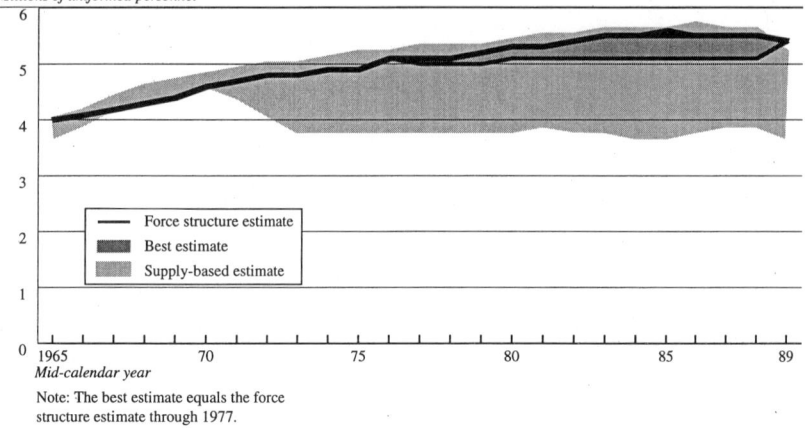

Figure A.1
**Best Estimate of Soviet Military Manpower, 1965–89**
*Millions of uniformed personnel*

Note: The best estimate equals the force structure estimate through 1977.

## Procurement Expenditures

Procurement expenditure estimates changed from one year's estimate to the next for three reasons: changes in the base year for either dollar or ruble prices, changes in the intelligence assessments of the size and timing of the production programs or technical characteristics of the procurement items, and changes in the estimates of costs of specific weapon systems and other equipment. In 1976–90, the base year for the dollar prices changed each year and accounted for a substantial part of the changes in the dollar estimates (e.g., almost all of the changes in the estimates published in January 1978 and February 1983 and two-thirds of the changes in the September 1979 and October 1981 estimates). The switch from a 1970 to a 1982 ruble price base in 1985 resulted in a large, onetime rise in the ruble estimate of procurement, but the ruble price base was not a factor in other years.

During most of this period changes in estimates of physical production did not significantly affect aggregate procurement spending.[8] There were numerous changes in individual categories of procurement, but they were mainly offsetting. The decision to include outlays for civil space programs in the 1980 ruble estimate raised procurement by about 4 billion rubles. In later estimates this figure was scaled back considerably, and civil space was dropped as an element of procurement in the 1989 estimate.

The 1989 estimate, however, included substantial reductions in the estimates of production for all years in the 1976–90 timeframe. These reductions reflected a new element of discipline introduced in the late 1980s within the intelligence community regarding estimates of physical production. Rankled by disagreements between CIA and DIA over many of these estimates, DDI Gates arranged in 1985 for an interagency reconciliation of differences in the physical estimates under the aegis of the NIC. The result was a series of interagency memorandums. The first appeared in March 1986 with the explanation, "This Memorandum establishes an interagency database on the yearly production of Soviet strategic and general purpose weapon systems and equipment for the period 1974–85."[9] The confrontation and justification of alternative estimates produced considerable convergence. In the great majority of cases interagency agreement was reached. Where it was not, the two agencies were required to explain their differing methodologies and the key assumptions underlying their estimates. The last in this series of memorandums was issued in September 1989. Although the effort involved a large number of analysts and an extended series of meetings, it was well worth the price. It improved the estimates by forcing consideration of the evidence and alternative analytic perspectives, and by requiring more thorough documentation of the estimates.

As the results of the external costing contracts were assimilated, changes were made in the dollar costs and, in most instances, the ruble costs of procurement items. In most of the contract work the costing methodologies permitted adjustments to the initially estimated cost when new technical information on a particular weapon or equipment item became available. The process, however, raised and lowered individual procurement prices, and the net effect on total procurement spending was generally small. After the 1976 major upward adjustment in ruble prices—and aside from changes in the base year—the overall dollar and ruble estimates of Soviet procurement spending were remarkably insensitive to new intelligence on prices. The drivers of the changes in the estimated value of procurement spending from estimate to estimate were changes in the physical production estimates.

## Operations and Maintenance Expenditures

The estimates of specific O&M costs were refined substantially during the late 1970s and the 1980s as attention turned to the exploitation of new evidence on Soviet operating practices. Table A.2 illustrates the most significant of these changes. With the exception of the Best-85 estimate

TABLE A.2

CHANGES IN ESTIMATES OF RUBLE OUTLAYS
FOR OPERATIONS AND MAINTENANCE[a]

| | |
|---|---|
| Comparisons | |
|   Best-79 and Best-76 | Average increase in 1976–79 was 12 percent |
|     Major changes | |
|       Up | Maintenance of aircraft **, electronic equipment *, general purpose vehicles *, military space system and airfields; utility costs *; and POL for vehicles and aircraft |
|       Down | Maintenance of personnel facilities * and missile systems *; transportation *; and civilian personnel * |
|   Historic Best-80 and Best-79 | Average increase in 1976–80 was about 5 percent |
|     Major changes | |
|       Up | Maintenance of ships and boats * and general purpose vehicles * |
|       Down | Maintenance of ground force equipment * and aircraft * |
|   Best-84 and Best-80 | |
|     Major changes | Average increase in 1976–80 was about 15 percent (includes effect of moving to a 1982 price base) |
|       Up | Maintenance of electronic equipment ** and missile systems *; utility costs * |
|       Down | Maintenance of nuclear weapons * |
|   Best-85 and Best-84 | Average increase in 1976–85 was about 50 percent (includes effect of moving to a 1982 price base) |
|     Major changes | |
|       Up | Maintenance of aircraft **, missiles **, ships and boats *, electronic equipment *, and ground force equipment; aircraft POL; preinduction military training; and civil space operations |
|       Down | No major revisions downward with conversion from 1970 to 1982 prices |
|   Best-88 and Best-85 | Average increase in 1976–85 was about 5 percent |
|     Major changes | |
|       Up | Maintenance of ground force equipment **, military space systems *, ships and boats, aircraft, and general support facilities; civil space operations * |

## TABLE A.2 CONTINUED

| Comparisons | | |
|---|---|---|
| | Down | Maintenance of missile systems * and electronic equipment *; utility costs * |

ᵃThe comparisons are of CIA best estimates of selected years. Major changes are defined as positive or negative changes that are large (in absolute terms) relative to the net overall change.
*Changes of 10 percent or more in relation to the net overall change
**Changes of substantially more than 10 percent

(which also incorporates the transition from a 1970 to a 1982 ruble price base) the changes in the individual estimates of total O&M were not all that large, although they all raised the O&M estimate.

With few exceptions, the changes in O&M estimates were occasioned by revisions in the estimates of equipment and weapons maintenance (for instance, ships and boats, aircraft), and not maintenance of facilities or housekeeping expenses such as utilities, transportation, POL, and preinduction military training. Broadly speaking, the O&M estimates after 1975 relied less on U.S. analogs and more on Soviet practices as laid out in open-source literature on maintenance of civilian vehicles, aircraft, and ships; defector reports on maintenance procedures applicable to particular weapons and equipment; technical collection on Soviet operations (flying hours for military aircraft, days at sea for naval ships and boats, kilometers driven by tanks, armored personnel carriers, and general purpose vehicles); and the Soviet military manuals mentioned earlier.

In the 1976 estimate, for example, the abundant information available on maintenance of civilian vehicles was applied to military trucks and cars, raising the estimate substantially. The following year, new maintenance cost factors were derived for all major land arms. They were based on life-cycle costs for major overhauls and periodic technical servicing as reported in Soviet sources. Like all maintenance estimates, the costs included outlays for parts and material for all levels of repair, as well as the labor and overhead charges for capital repair and modernization. (Routine repair was carried out by armed forces personnel whose wages were subsumed under personnel costs.)

Major studies of Soviet practice in maintaining aircraft and ships and boats, however, provided the largest increases in O&M estimates.[10] In-

Costing Improvements

stead of U.S.-based cost factors, the aircraft maintenance paper used factors uncovered in Soviet sources. They raised the estimate of aircraft maintenance costs by 4–5 billion rubles (1970 prices) and produced a higher level of costs and a faster rate of growth in spending than the previous methodology.[11] The aircraft study found that Soviet maintenance was more expensive than U.S. practice because of "extremely conservative Soviet maintenance norms and the short service lives associated with the Soviet airframes and engines."[12] The paper on naval ships and boats concluded that the Soviet navy planned maintenance by applying percentage factors to the cost of each ship in the fleet. Several unclassified Soviet publications on maintenance of merchant ships supplied the basis for the cost factors, and the applicability of the approach to the Soviet navy was confirmed by a variety of human sources. As a result, the estimated cost of maintaining Soviet naval ships increased by more than 2 billion (1970) rubles in the 1980 estimates.[13] And, as in the case of aircraft, the new methodology pushed up the estimated rate of growth of ship maintenance. In the old methodology, maintenance costs were based on tonnage. Thus a modern, more expensive destroyer was assigned the same maintenance cost as an older, less sophisticated destroyer of the same displacement. Basing maintenance cost factors on the cost of the ship rather than displacement reflected changes in complexity.

The O&M estimates were revisited periodically during the 1980s. In the 1987 estimate, the estimates of land arms maintenance were again revised to take advantage of a new intelligence community database on the size and composition of Soviet ground forces. The database tracked changes in force readiness, which in turn dictated changes in operating and maintenance cost factors.[14]

SOVA analysts recognized that intensified economic stringencies appeared to have curbed the growth of procurement and looked for evidence that economies had been introduced in O&M as well. A decline in naval ships' days at sea was noted in the mid-1980s, but it was more difficult to detect departures from the planned maintenance procedures that had become the basis of SOVA's O&M estimates. It is likely that pressures to save resources that increased in the Gorbachev years did affect O&M, but in ways that SOVA was unable to quantify and reflect in revised cost estimates. Indeed, a 1988 SOVA article reported "mounting evidence that the Soviet military has begun to revise training procedures, curb operating rates, and extend the service lives of equipment."[15]

In addressing the problem of O&M costs, SOVA analysts learned a good deal about military readiness as it related to maintenance of weapons and equipment. The Soviet armed forces were found to rely on regular preventive maintenance, replacement rather than repair of defective components and parts, periodic rebuilding of entire airframes and engines, specialized maintenance personnel, and conservative—that is, frequent—scheduling of maintenance. The pervasive Soviet use of conservative planning factors, or "norms," was expensive. It required large stocks of spare parts and weapons. The frequent overhaul of aircraft and tank engines pushed up the demand for spares. On the other hand, Soviet forces tended to keep the newest weapons and equipment in reserve while using older equipment in exercises. In the initial stages of a conflict they could expect a higher level of reliability than if they began a war with the equipment they had been using for some time.

## New Estimates of Construction Costs

As CIA work on O&M increasingly depended on Soviet planning factors and the Soviet planners' preference for standardization, so did its research on military construction. Analysts saw a shift in the orientation of the USSR's defense construction. Whereas in the 1950s and 1960s the emphasis was on facilities supporting the strategic forces (airfields and missile complexes), attention turned to ground and naval bases and, in the 1970s, to the logistical infrastructure.

In the 1979 estimate the coverage of military construction in the CIA estimates was expanded to include off-base family housing, tank-turret strong points along the USSR's borders, the cost of facilities related to premilitary training, and overseas bases.[16] By this time, however, OSR realized that an attempt to identify and cost every military facility would be expensive and unlikely to succeed. As new facilities were discovered in 1976–78, construction cost estimates climbed. Working with the Office of Imagery Analysis, OSR developed a sampling technique to ascertain the construction histories of typical units and prorate the costs of this construction over time.[17] In the 1980s, then, construction cost estimates were based on a mixture of direct observation of individual building projects (for example, airfields) and the derivation of per-man factors for unit-associated facilities that tended to be standardized over the whole Soviet Union (for example, those serving ground force divisions and missile complexes). The result was a doubling of the construction estimate in rubles in 1981.[18] The equivalent dollar estimate increased in proportion because this was one of the few areas where the initial estimate was made in rubles and then translated to dollars using a weighted-average ruble-to-dollar ratio for construction that had been developed in OER in 1976.[19]

The completeness of the construction estimate's coverage remained an issue. SOVA told the Selin panel in late 1982 that its construction estimate probably was not systematically biased upward or downward (although there were probably significant estimation errors) "because we don't cost what we see, we cost what we estimate to be there."[20] At the same time SOVA conceded that the construction estimate did not try to include the cost of the major infrastructure (road construction, electric power lines) that supported the military.[21] But the SOVA construction estimate did include the cost of specific underground facilities for leadership protection, military command and control, and weapons storage as they were identified. As of 1988 the estimated value of these underground construction activities amounted to about $34 billion (in 1986 dollars) for 1950–88. Seventy percent of this construction was said to have been carried out during 1945–65.[22]

A related issue, the size and nature of the Soviet civil defense program, came to the forefront in the mid–1970s. The "Team B" group of outside experts commissioned by DCI Bush to review the NIE's dealing with Soviet strategic programs had criticized them for displaying a lack of interest in civil defense, even though there was evidence of an upswing in civil defense activity after 1971.[23] In response the intelligence community looked at the question again and concluded that "Soviet civil defense programs are steadily improving. The program is more extensive and better developed than we had previously understood."[24] The estimate described the program as "continuing, steady," and not a "crash effort."[25]

Lee has advanced an alternative view of the scope of construction related to command control and some part of civil defense: "From relatively low levels in the 1950s and most of the 1960s, the cost of constructing bunkers and deep underground structures for the nuclear war management system evidently rose to about 10 billion rubles annually circa 1970, and soon rose to some 20 billion per year."[26] After taking account of all of the new evidence, CIA's 1990 estimate of total military construction for the 1980s never exceeded 5.5 billion current rubles in any year.

## A New Approach to Military RDT&E

Along with the extensive revisions of the ruble estimates of military procurement and O&M, OSR moved to improve the RDT&E estimate. The director of OSR forwarded a study proposal to the DDI in January 1976 noting the widespread criticism of the existing methodology and declaring it "imperative, therefore, that new research and analysis be undertaken."[27] Later that year a CIA study group on military RDT&E was established. It toured the country scouting for suggestions for upgrading the estimate. Meanwhile, the estimate was corrected from year to year to take account of new open-source information and, more importantly, new understandings of the definitions underlying published Soviet figures. External contract work also targeted the dollar-equivalent costs of parts of the USSR's R&D. For example, a 1981 study estimated that all Soviet strategic ballistic missile development programs from 1960 to 1980 would have cost the United States $75 billion.[28]

The search for an alternative approach to the RDT&E estimate first focused on indicators of change in R&D programs and the development of a microlevel database on the Soviet R&D establishment. A 1982 report tracked growth in the number of weapon development programs and the physical expansion at R&D facilities and concluded that military R&D would continue to rise.[29] A report the following year further developed this theme and reported that CIA's military RDT&E installation file now included about nine hundred research institutes, design bureaus, and test facilities.[30]

Work on a way of distilling an estimate of the cost of Soviet military RDT&E from collected microinformation continued in 1984–85, and in 1986 a published report marked the incorporation of a new methodology in CIA's estimate of Soviet defense spending.[31] The approach is sketched in figure A.2. In the first stage, facilities were identified, and where possible their purpose and staffing were established. Imagery analysis provided estimates of the floor space of the facilities. Research found 1,100 facilities connected with military RDT&E after 1965. Because it was clear that the facilities file did not include all Soviet military RDT&E establishments, a statistical technique known as "capture-recapture" (frequently employed in wildlife management) was used to estimate the number of missing facilities.[32] The final estimate was 1,500 military R&D facilities.

When the purpose, staffing, and floorspace of facilities were known, the cost of RDT&E work was estimated by multiplying manpower and floorspace estimates by cost factors based on open and classified sources.[33]

**Figure A.2**
**The Process of Estimating the Resource Costs of Soviet Military RDT&E**

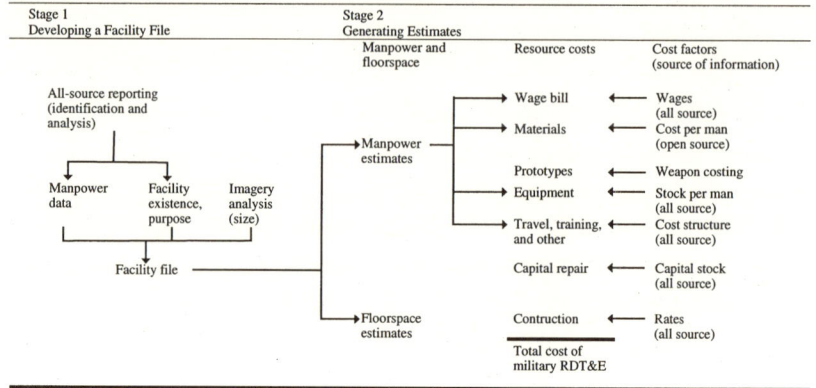

**Figure A.3**
**Estimated Soviet Military RDT&E Expenditures, 1965–84**

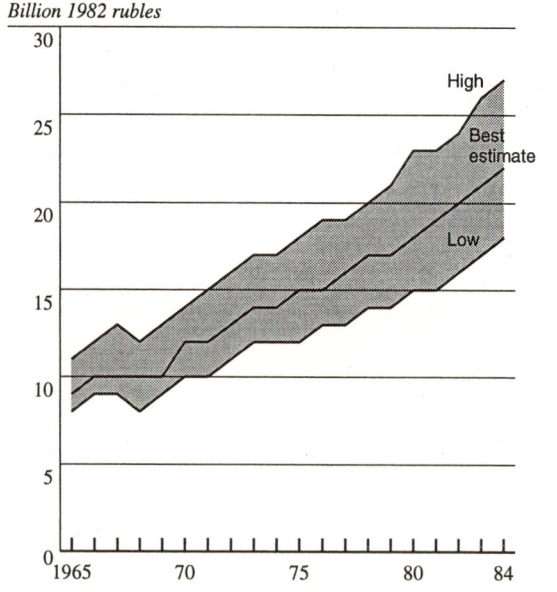

The 90 percent confidence interval in the chart averages +20/-15 percent of our best estimate for each year.

The costs for different kinds of RDT&E facilities were then extrapolated to all 1,500 facilities in the file. The results of the intensive R&D effort directed at Soviet military RDT&E are depicted in figure A.3. The best estimate of these outlays (in the 1982 constant prices then serving as the base for CIA estimates) had them rising from 9 billion rubles in 1965 to 22 billion rubles in 1984—an average annual increase of 4.8 percent. Using the old methodology, military RDT&E grew by 6.8 percent annually (in 1970 prices). The range of uncertainty was calculated in a Monte Carlo simulation of the effect of uncertainties attached to the various stages of the estimation process. The new procedure also gave an estimate for total Soviet RDT&E outlays that agreed fairly well with figures revealed by Soviets in prominent scientific positions or engaged in research on R&D (figure A.4). During the 1965–84 period the USSR devoted roughly three-fifths of its total RDT&E spending to defense research, according to the estimate.

Translating the ruble estimate into dollars was a thorny assignment.

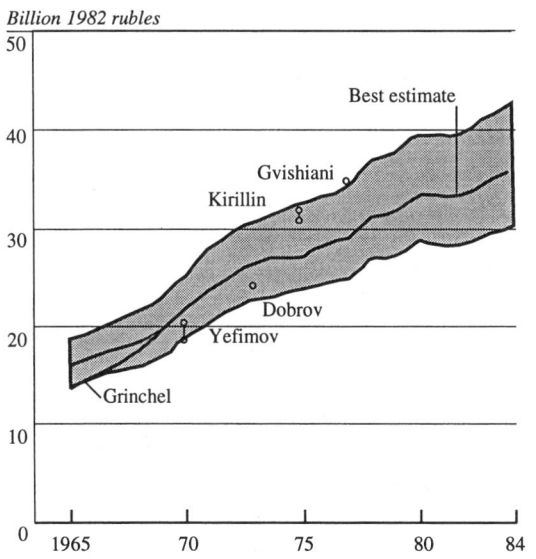

**Figure A.4**
**Soviet Total RDT&E Expenditures, 1965–84:**
**CIA Estimate Compared With Soviet Statements**

The 90 percent confidence interval in the chart averages +20/-15 percent of our best estimate for each year.

Costing Improvements

The objective was to measure what it would cost the United States to replicate, with U.S. efficiency or productivity in the use of R&D resources, Soviet RDT&E results. As a proxy for the necessary RDT&E ruble-to-dollar ratio, SOVA decided to apply the average of the ruble-to-dollar ratios for major Soviet weapons, reasoning that it was a good approximation of relative U.S. and Soviet efficiencies in the later stages of weapon engineering development and prototype production—the most expensive stages of RDT&E. Like the ruble estimate, the new estimate also gave a lower figure for growth in the dollar cost of Soviet RDT&E during the 1965–84 period: 4.4 percent per year compared with 6.8 percent per year in the old estimate.

The problem with the new approach to the estimation of defense RDT&E was keeping it current. The methodology required a great deal of microinformation and was reliable for the 1970s and early 1980s where most of the data points were found. New information had to be collected and could be reflected in the estimates only with a lag of a few years. In the late 1980s SOVA adjusted the estimates to take account of Soviet claims that defense industry and its research establishments would be asked to do more for civil industry. Still, CIA's military RDT&E estimate first declined in 1990, when it fell by 5 percent compared with the USSR's announced intention to cut it by 14 percent.

# Appendix B

## The Index Number Problem and Defense Programs: Some Considerations

The essence of the objection raised by some economists to the dollar concept used by CIA to size the Soviet defense effort is that, according to economic price theory, the CIA approach exaggerates the size of the Soviet effort. The objection is not that the CIA overprices the individual elements of the Soviet program, but that the Soviet program itself would be different and cheaper in dollars if the Soviet defense managers had faced the U.S. structure of relative prices rather than the Soviet structure of relative prices. This conclusion follows from the fundamental precept of price theory that a purchaser will strive to minimize costs by substituting less expensive inputs for more expensive inputs in achieving a given level of utility (in this case, defense capability) and from the empirical fact that no two national price structures are identical.

A concrete example may help to clarify the theoretical proposition. Suppose that the manager of a U.S. military depot finds that he or she needs more capability to move material around in a large warehouse. If one fork-lift truck does the work of twenty people but only costs what five additional people would cost, additional equipment rather than people will be emphasized to solve the problem. However, the outcome might be markedly different at a Soviet depot with the same problem. If the Soviet forklift also does the work of twenty people, but costs the equivalent of thirty additional people, then people rather than equipment will be emphasized.

Now when the actual Soviet solution to the problem is expressed in dollars using the U.S. price structure—as CIA estimates the dollar prices of Soviet defense activities—it will be more expensive than the actual U.S. solution, because manpower, which is relatively costly in the United States, will weigh more heavily in the calculation. The converse is equally true. If both solutions were expressed in rubles reflecting the Soviet structure of relative prices, the U.S. solution would be more expensive than the Soviet solution, because equipment, which is relatively more expensive in the Soviet Union, would weigh more heavily in the calculation of the U.S. program. In fact, both countries have purchased an equal amount of additional logistic capability. But when measured in dollars it appears that the Soviet Union has purchased more than the United States; and when measured in rubles it appears that the United States has purchased more.

This conundrum in international monetary measurement—which applies in exactly the same way to intertemporal monetary measurement within a country—is identified in the economic literature as the "index number problem." Given the frequent requirement in economic affairs to make international comparisons, economists have adopted the general convention of either (1) presenting each country's activity valued in its own prices and in the prices of the other countries involved in the comparison or (2) displaying the bilateral relationship between any two countries as calculated as an average (the geometric mean) of the two relationships arising from first comparing the two countries' activities expressed in one country's prices and then comparing the two countries' activities in the other country's prices.

The question then is, Why did not CIA's military-economic analysts employ the conventional approach of international economists in their effort to size the Soviet defense programs and compare the results with the counterpart U.S. programs? There are both theoretical and practical components in the answer.

Critical assumptions underpinning the theory of the index number problem as outlined in the example above are that purchasers of defense goods and services possess perfect knowledge of available alternatives and that market mechanisms (including price flexibility) work perfectly to reflect that knowledge. Market imperfections in the defense sector on the U.S. side are real, but insignificant enough to ignore. The same, however, cannot be said for the Soviet side. Deviations from perfect market solutions are likely to be far greater so that applying the Soviet price structure to the U.S. programs is likely to introduce a more significant source of error than applying the U.S. price structure to Soviet programs.

A more important theoretical objection that the military-economic analysts had to using the conventional approach was that it implied just the

Appendix B

kind of judgments about military capability/effectiveness that they were trying to avoid. Notice how intrinsic the words "utility" and "capability" are to the explanation of index number theory presented above. To the CIA military-economic analysts there was no way that the conventional approach of the economists could capture the subtle, intangible, and subjective aspects of military capabilities and effectiveness that would be implicit in using the theory. It was the view of these analysts that it was far better to apply the U.S. price structure to Soviet programs and then frankly point out that the calculation was limited to costing, in dollars, the *inputs* to the Soviet program and that no attempt was being made to evaluate the output—that is, the military capabilities/effectiveness that the inputs produced.

Another consideration that influenced the CIA military-economic analysts was a combination of the index number theory involved and empirical observation. In the defense sector, which is highly dependent on the most advanced technology, the ability to substitute goods and services based solely on relative prices would be severely constrained and, as a consequence, the distorting effects of the index number problem in the defense sector would be far less than in the economy as a whole. While it might have been possible for the Soviets to substitute relatively cheap unskilled manpower for relatively expensive fork-lift trucks—as in the military depot example postulated above—the opportunities for such substitutions are generally limited by the demands of modern military technology. Advanced weapons systems such as ICBMs simply can not be designed, built, and deployed without using the most scarce resources.

Even if CIA's military-economic analysts had decided that the conventional approach was appropriate, practical considerations made it almost impossible to pursue effectively. As this book amply demonstrates, finding reliable ruble prices for Soviet military goods and services was a daunting intelligence collection and analytic task that was not consistently successful. Establishing meaningful ruble prices for U.S. military programs would have been far more challenging. For most of the Cold War and for most areas of military technology, the United States was ahead of the Soviet Union. This would have meant that the CIA military-economic analysts would have had to establish valid ruble prices for items that the Soviets at the time could not even build! Not a research task that held much promise for success.

Finally, it must be borne in mind that the CIA military-economic analysts approached their task first and foremost as an intelligence problem and not as an academic exercise in theoretical economics. Their goal as intelligence officers was to present to the U.S. national security community in credible and easily understood terms the best estimate they could make of the size of the Soviet military effort.

# Notes

## Chapter 1. Introduction

1. An elaborate and generally unsuccessful attempt to evaluate and convey the magnitude and trend of Soviet military programs without the use of a common denominator was made in the mid-1980s.
2. See chap. 6 for further discussion of the index number problem.
3. This is more than a minor academic consideration. A frequent criticism of CIA's spending estimates over the years was that they did not grow fast enough to accommodate new weapons programs coming online. Often critics failed to perceive or ignored the offsetting programs going offline. In one important sense the focus on new programs is understandable. They pose the threat—they can kill you. Things that are not occurring are not a threat.
4. After World War II only figures for total military spending were included in the Minister of Finance annual budget announcements. The lack of supporting detail and other factors raised many analytic questions about the coverage of the announced figure. The custom of providing no details prevailed until 1989, when Mikhail Gorbachev released a more detailed breakdown of defense spending.

## Chapter 2. Getting Started, 1950–60

1. National Security Council Intelligence Directive (NSCID) no. 3, Jan. 13, 1948, provided the following delineation of dominant interests by agency: Department of State: Political, Cultural, and Sociological Intelligence; Department of Army: Military Intelligence; Department of Navy: Naval Intelligence; Department of Air Force: Air Intelligence; each agency (including CIA) in accordance with its respective needs: Economic, Scientific, and Technological Intelligence.
2. Dave Coffin, *Development of Economic Intelligence,* vol. I, *1950–1960,* p. 3.

3. DCID 15/1, *Production and Coordination of Foreign Economic Intelligence,* Sept. 14, 1954.
4. Ibid. See also the updated version of DCID 15/1, June 10, 1958, which makes the authority more explicit by identifying "economic cost studies of Bloc military programs" as a legitimate subject of intelligence production by CIA.
5. This method was independent only in the sense that it rejected the announced defense budget as a reliable measure. It continued to rely, however, on the validity of other Soviet economic statistics.
6. We discuss some of these nonintelligence estimating efforts in chap. 6.
7. At the time even a soldier's pay was treated as a state secret, and nearly all industrial activity was considered off limits to westerners.
8. CIA, *The Role of ORR in Economic Intelligence,* pp. 1–3.
9. See the section on monetary measures—rubles and dollars—in this chapter for a discussion of the different conceptual objectives for estimating the dollar values of Soviet defense activities and estimating the Soviet ruble outlays for defense.
10. Analytic interest in the institutional (i.e., military service) breakdown of spending shown in table 2.2 was largely supplanted in the early 1960s by focus on the functional (i.e., military mission) breakdown shown in table 2.3, but it reemerged in the 1970s.
11. The late former director of the Defense Intelligence Agency, Lt. Gen. (ret.) Daniel Graham, for example, for many years dismissed CIA estimates of total Soviet spending as being far too low; he believed CIA analysts were guilty of incorrectly applying the analogy of the bureaucratic give-and-take of the U.S. defense budget process to the Soviet defense budget process ("mirror imaging") in order to place an artificial limit on the total budget's size. The analogy of the U.S. defense budget process cited by Graham—most recently at a conference at Harvard University in Dec. 1994—was not employed in any way by CIA analysts to estimate Soviet military spending. The levels and patterns of CIA spending estimates since the mid-1950s have been driven by the underlying physical estimates of Soviet military forces and programs made by the U.S. intelligence community, not by judgments about how the Soviet budget process worked.
12. Soviet military trade prices for such items were more available than internal prices, but they were generally found to be unreliable proxies for internal prices because they were influenced by foreign policy objectives and international competition.
13. Over the years the intelligence community acquired an impressive inventory of such material. Several pilots, Soviet and otherwise, defected to the West along with their Soviet-made military aircraft. Unfortunately none of this material came complete with sticker price.
14. CIA/DI, DIA, *Dollar Costing of Foreign Defense Activities: A Primer on Methodology and Use of the Data.*

15. For a more extensive and technical discussion, see CIA/DI, *A Guide to Monetary Measures of Soviet Defense Activities.*
16. In fact the CNO could spend the $20 billion more efficiently on a different mix of naval programs.
17. CIA/DI, DIA, *Dollar Costing*, p. v.
18. In one form or another CIA publications carefully explained the meaning of the dollar values of Soviet defense activity. That misunderstanding persisted despite these explanations suggests that the explanations were ignored, CIA did not do a good job making the point, or some who seemed to misunderstand had an agenda making it convenient to misunderstand.
19. For a discussion of monetary measures of defense and the variety of questions they address, see CIA/DI, *Monetary Measures.*
20. For example, see Abraham S. Becker, *Prices of Producers' Durables in the U.S. and USSR in 1955.*
21. Given available information, it is not possible to assess the validity of the 1955 dollar-to-ruble conversion ratios developed and used by CIA military-economic analysts. An examination of analyst files and published papers of the period indicates that few actual Soviet ruble prices of military equipment were known, and that known ruble prices of civilian equipment played an important role in establishing dollar-to-ruble conversion ratios for military equipment. It was generally believed that Soviet military industry was considerably more efficient than civilian industry, and this judgment probably influenced the construction of the dollar-to-ruble ratios for military equipment. Subsequent events suggest that if the estimated ratios for 1955 were in error, such error was probably in the direction of being too high, in effect overstating the efficiency of Soviet military industry relative to that of the United States. See chap. 4.
22. See *Sherman Kent and the Board of National Estimates, Collected Essays,* ed. Donald Steury, pp. x–xii, for a brief review of some of the period's key bureaucratic developments.
23. According to Gen. Andrew Goodpaster, secretary of the White House Staff under President Dwight D. Eisenhower, the president expressed the view that "there was a lot of self-interest in the intelligence assessments of the military services—they were out to promote their own programs." John L. Helgerson, *CIA Briefings of Presidential Candidates, 1951–1992,* p. 40.
24. For example, see Lawrence Freeman, *U.S. Intelligence and the Soviet Strategic Threat,* pp. 183–98; Fred Kaplan, *The Wizards of Armageddon,* pp. 155–73; John Prados, *The Soviet Estimate,* pp. 49–50.
25. While the motives for ORR's entry into military intelligence were mixed, that it did so under the aegis of its military-economic responsibilities is clear.
26. This does not mean that CIA analysts were always right and DoD analysts were always wrong, that DoD analysts generally lacked intellectual integrity, or that CIA analysts were miraculously free of bias. The point is

27. NIE-3, *Soviet Capabilities and Intentions*, p. 3.
28. NIE-64, *Soviet Bloc Capabilities through Mid-1953*, part 1, p. 3.
29. NIE-65, *Soviet Bloc Capabilities through 1957*, p. 7.
30. NIE-11-4-54, *Soviet Capabilities and Probable Courses of Action through Mid-1959*, pp. 1–2.
31. NIE 11-6-54, *Soviet Capabilities and Probable Programs in the Guided Missile Field*, p. 2.
32. CIA/ORR, "Contribution to NIE 11-5-55, Air Defense Capabilities of the Sino-Soviet Bloc, 1955–1960."
33. Ibid., p. 26.
34. NIE 11-5-55, *Bloc Air Defense Capabilities, 1955–1960*.
35. From an interview of John G. Godaire conducted in June 1970 as background for an internal history project that was never published.
36. Formally named the IAC [Intelligence Advisory Committee] Ad Hoc Military Cost Committee. In addition to the CIA chair it included members from the Departments of Air Force, Army, Defense (Joint Staff), Navy, and State.
37. At least some in the service intelligence units saw the costing work as a potential infringement on their heretofore exclusive prerogative for quantifying and describing "their piece" of the Soviet threat.
38. NIE 11-4-56, *Soviet Capabilities and Probable Courses of Action through 1961*, pp. 75–80.
39. Ibid., pp. 14, 16–17.
40. Ibid., p. 75.
41. CIA/ORR, *Military-Economic Programs of the USSR, 1947–1961*, p. 1.
42. NIE 11-4-56, *Soviet Capabilities*, p. 75.
43. The issuance of postmortem memoranda after the completion of major NIEs, in part to provide guidance for preparation of the next NIE on the same subject, was a standard ONE procedure at the time. This memorandum on NIE 11-4-56 is undated.
44. The following titles illustrate the range and nature of the CIA/ORR research effort: *Theory and Application of the Learning Curve; Estimates of Costs in Rubles of Building Various Types of Merchant Ships in the USSR; Pay for Personnel of the Soviet Ground Forces, 1957; Costs of Construction of the Soviet W-Class Submarine; 1955 Ruble-Dollar Ratio for Cost of Construction of the Soviet "SKORYY" Class Destroyer;* and *Ruble-Dollar Ratio for Soviet Aircraft*.
45. CIA/ORR, *Soviet Military Expenditures, 1958–1965*, pp. 20–23.

## Chapter 3. New Questions, New Techniques, and New Evidence, 1961–74

1. Kaplan, *Wizards*.
2. See chap. 7 for an example of how CIA's estimates of Soviet defense

costs played a key role in a major U.S. defense decision in the early 1960s.
3. A. W. Marshall, J. E. Loftus, and G. E. Pugh, *Project Lamp, Systems Analysis and the Military Estimates Process*.
4. Ibid., p. 1.
5. Sherman Kent, Memorandum for Deputy Director (Intelligence), Jan. 10, 1963.
6. The name was changed because DoD planners did not wish to use a document labeled "Assumptions" as a primary basis for their planning. Several alternative names were proposed and "Projections" was selected as the most acceptable to the producers and primary customers in DoD. The name change did not affect the content of the document.
7. Cyrus Vance to the Honorable John A. McCone, Director, Central Intelligence Agency, Feb. 5, 1965.
8. J. G. Godaire, "The Claim of the Soviet Military Establishment," in *Dimensions of Soviet Economic Power*, pp. 35–46.
9. *Allocation of Resources in the Soviet Union and China* (1974).
10. *Allocation of Resources in the Soviet Union and China, 1983*, p. 1. Senator Proximire added: "The views of the spokesman for the intelligence agencies have often provoked comments from others in public and private circles and have sparked discussions in Congress and the media. The published writings have been used in innumerable scholarly and popular writings, the hearings have thus achieved one of our primary objectives: namely, broadening and increasing the quality of the public dialogue about the Soviet and Chinese economies."
11. CIA/OSR, Memorandum for the Record, "Transmittal of Information to Congressman Aspin." Donald Burton, Chief of the Programs Analysis Division in OSR, told Aspin that Soviet defense spending was currently in an expanding phase with growth averaging 5 percent a year, but when ICBM deployment programs were completed "Soviet spending will level off at a new plateau."
12. George Cary to the Honorable Les Aspin, Nov. 28, 1975.
13. Donald F. Burton, "Activities of Military-Economic Planning Branch in Fiscal Year 1970."
14. Edward W. Proctor, Memorandum for Acting Assistant Director, ORR, June 18, 1964.
15. The two large military spending papers—one broken down by military mission and the other by economic resource category—cited at the end of chap. 2 are the primary examples.
16. The other titles in this series are CIA/DI, *The 1970 Soviet Defense Budget in Perspective: Trends in Spending for Defense and Space Since 1960;* CIA/DI, *Soviet Expenditures for Defense and Space Programs, 1962–1971;* CIA/DI, *Soviet Defense Expenditures, 1963–1972*.
17. CIA/DDI, Memorandum for the Director of Central Intelligence, "Costing Military Programs," Apr. 25, 1972.

18. Ibid. Edward Proctor and Bruce Clarke, director of OSR, believed it was particularly important that panel members have full access to the same information available to military-economic analysts so that the panel could understand the nature of the evidentiary base of the building-block method. The concerns seem justified by the fact that although DIA subsequently convened a conference of experts, it failed to produce any tangible impact on the estimating process.
19. Richard Helms, Memorandum for Lt. Gen. Donald V. Bennett, Apr. 25, 1972.
20. Other members of the first MEAP were Dr. Herbert Levine, University of Pennsylvania; Dr. William Niskanen, University of California; Dr. K. Wayne Smith, vice president of Dart Industries; and James Plummer, vice president and general manager of Lockheed Corporation's Missile and Space Center. Among those who later served on MEAP were Lee Badgett, Abraham S. Becker, Robert W. Campbell, Kenneth W. Dam, Richard E. Ericson, Ed Hewett, Gail Lapidus, Stephen Meyer, Wesley Posvar, Thomas C. Schelling, Ivan Selin, Vladimir G. Treml, and Stanley Weiss.
21. CIA/Military-Economic Advisory Panel, *First Report—Military Economic Advisory Panel*. It should be noted that DIA military-economic analysts were regularly invited to attend any meetings of the panel to which CIA military-economic analysts had been invited.
22. Ibid., p. 3.
23. Ibid., p. 7.
24. The computerized system was initially identified as the MEBORR system, an abbreviation for the Military Expenditures Branch, Office of Research and Reports. A few years later the decision was made to identify the system by a functional rather than organizational acronym. As an aside it is interesting to note that the word "scam"—with its current unsavory connotation—was not part of the common lexicon at the time. Some of CIA's critics have been unable to resist the temptation to link the SCAM acronym to the current word.
25. As announced in mid-1967, the revisions of enterprise wholesale prices were concentrated in heavy industry, with sharp increases for branches producing raw materials (especially iron ore and coal) and smaller increases for products of heavy machinery branches where raw material accounted for a large part of total costs. Prices were reduced on average in machine-building branches where material costs were less important (instruments, radio equipment, and electronics). The State Price Committee reported, however, that average prices on all machine-building products had not changed.
26. CIA/OSR, Memorandum, "Price Base for Military Expenditure Series."
27. Ibid.
28. The ruble prices obtained were rarely unambiguous in important re-

spects—for example, whether the price of an aircraft included spare engines and/or other spares, whether it was the initial unit cost, the average cost for a production run, the marginal cost at some specific point on a learning curve, and the year for which the price is valid. The degree to which questions like these could be reliably answered significantly influenced the utility of the price information that was obtained.

29. Interview conducted by authors with James Schlesinger, Mar. 9, 1995.
30. The two supervisors of the burden paper had reported to the DDI: "We aim for a more extensive discussion of burden as we see it and as the Soviets might view it. Still, we are not confident that it will satisfy A. W. Marshall's hopes for a break-through in measurement." At the time Marshall was director of net assessments in the Pentagon and served as pointman for Secretary Schlesinger on Soviet military-economic matters. CIA/OSR and OER, Memorandum for Deputy Director of Intelligence, "Paper on the USSR's Defense Burden."
31. Daniel Graham, Memorandum for Deputy Director, Central Intelligence Agency, "Comments on CIA Paper, 'The Measurement and Meaning of Defense Burden in the Soviet Setting'."
32. Edward Proctor, Memorandum for Lieutenant General Daniel O. Graham, USA, Director, Defense Intelligence Agency, "Your Comments on 'The Measurement and Meaning of Defense Burden in the Soviet Setting'."
33. CIA, *The Economic Impact of Soviet Military Spending.*
34. It should be noted, however, that much of the work on dollar costs improved the dollar and ruble spending estimates. Although better dollar personnel costs had no impact on the ruble estimates because ruble personnel cost factors were applied directly to the Soviet manpower levels, ruble estimates of procurement spending were obtained by converting estimated dollar costs of Soviet weapons to rubles with a dollar-to-ruble conversion factor. The improvements in dollar costs of Soviet weapons obtained during this period, therefore, affected both the dollar and ruble estimates of procurement spending.

## Chapter 4. The Estimates, 1975–90: Substantial Progress and Managerial Doubts

1. CIA, *Estimated Defense Spending in Rubles,* and CIA/DI, *Estimated Soviet Defense Spending in Rubles, 1970–1975.*
2. The terms Best-75 and Best-77 are identification labels used to distinguish between different expenditure estimates. The 75 and 77 identify the years in which the estimates were made and the term Best indicates that the expenditure estimates were designed to reflect the underlying quantitative data on forces and programs that were considered by CIA analysts to be most likely. In addition to these best estimates of expenditures others were made for many purposes. For example, if the NSC wanted to know in 1975

how much the Soviets would save under various arms control scenarios, estimates would be made and labeled something like Arms Control/1-75, Arms Control/2-75, etc.

3. Burton testimony before the Methodology Panel of the Working Group on Soviet Military Economic Analysis (hereafter the Becker Panel), Nov. 20, 1982, pp. 18–21, and Burton, "Estimating Soviet Defense Spending," *Problems of Communism* (Mar.–Apr. 1983): 87–88.
4. Burton testimony, p. 21. Ruble-to-dollar ratios for missiles and aircraft, however, were raised in 1974.
5. CIA/DDI, Memorandum from Deputy Director for Intelligence, "Meeting with OSR and OER People, Friday 24 October 1975, 1530 hours, re: Soviet Emigré."
6. Burton testimony, p. 21. Burton told the Becker panel that the emigré "gave us an excuse for stopping for a year to assimilate all [of these new prices]."
7. When the joint Office of Economic Research-Office of Strategic research paper on the defense burden was published, Schlesinger was skeptical and expressed his reservations to a group of agency representatives who met to discuss his problems with the paper.
8. Committee on the Present Danger, *Russian Military Expenditures*, p. 4. The Committee added that the "Team B" exercise had forced CIA to "double their 1970 estimate for Soviet military spending" and "to nearly quadruple their procurement estimates." In fact, Team B's findings had nothing to do with the revision. The actual revision was much smaller than the committee asserted (see table 4.1).
9. The controversy surfaced again in 1991 in CIA's *Studies in Intelligence*. Norbert Michaud, a DIA analyst who had participated in debriefing the emigré, wrote an article, "The Emigré Who Raised the Burden," in which he said CIA was reluctant to accept the emigré's story and tried to undermine his credibility. Michaud also asserted that before the emigré's appearance, the ruble prices for procurement were not being reviewed: "Any information suggesting that prices were in fact higher was dismissed" (p. 59). Noel Firth, acting director of the Office of Strategic Research during most of 1975–76, replied to Michaud's article in the winter 1991 issue of *Studies in Intelligence*. Firth described the lengthy debriefing of the emigré as essential to establishing his credibility. In this connection, he insisted: "Contrary to the suggestion in the article, CIA did not manipulate the process in some way to cause the source to fail his first polygraph interview. Indeed, by that time we were pretty well convinced that he was basically telling the truth, and we were as surprised and dismayed as anyone involved when he indicated deception in the initial interview" (p. 84). Firth also rejected Michaud's claim regarding the ruble prices for procurement, writing that long before the emigré arrived CIA was working to update the prices and had become convinced that they were too low. Norbert Michaud, "The Emigré Who Raised the

Burden," *Studies in Intelligence* 35, 2 (Summer 1991): 57–61; Noel E. Firth, "Soviet Defense Spending Controversy," *Studies in Intelligence* 35, 4 (Winter 1991): 83–87.
10. Steven Rosefielde, "On the Interpretation of Soviet Arms Procurement Expenditures under Conditions of Rapid Technical Progress," *Osteuropa Wirtschaft*, 25, 1 (1980): 42.
11. With an overall rise in the estimate of procurement in 1970 (in 1970 rubles) of 43 percent, land arms procurement increased by 33 percent, aircraft by 92 percent, naval ships and boats by 53 percent, electronic equipment by 23 percent, and nuclear weapons by 13 percent. The estimate for missile systems declined by 11 percent. These changes were driven by new price estimates, not the emigré's information, which did not contain such a breakdown.
12. George Bush, "Statement of Hon. George Bush, Director, Central Intelligence Agency," p. 4. He said he was "impressed with the Intelligence Community's constant reexamining of old judgments in the light of the unceasing flow of new information" and that he had told those responsible for the estimates that intelligence "ought to be prepared from the best information possible without partisanship, without fear of bias."
13. *Business Week*, Feb. 28, 1977, pp. 96–103.
14. *Washington Post*, Mar. 7, 1977, p. A21.
15. Eugene V. Rostow, "Foreword," *Understanding the Soviet Military Threat*, p. 3.
16. Cited in Franklyn D. Holzman, "Politics and Guesswork: CIA and DIA Estimates of Soviet Military Spending," *International Security* 14, 2 (Fall 1989): 106.
17. Testimony before the Consumer Panel of the Working Group on Soviet Military Economic Analysis (hereafter the Selin Panel), Jan. 13, 1983, p. 3. With understatement Chu characterized the current climate as one of "vague unease."
18. Testimony before Selin panel, Feb. 24, 1983, p. 7.
19. Franklyn D. Holzman, "Are the Soviets Really Outspending the U.S. on Defense?" *International Security* 4, 4 (Spring 1980): 97. He said the revision of 1976 "should be cause for rejoicing rather than for Cassandra-like warnings, since it indicates that in the military arms race, the Soviets had been forced to work much harder than had been imagined."
20. Holzman, "Are the Soviets Really Outspending," p. 98.
21. Philip Hanson, "Estimating Soviet Defense Expenditure," *Soviet Studies* 30, 3 (July 1978): 409–10. According to Hanson, even if the new burden measures were correct, one could not be confident that they reflected the true opportunity cost of defense spending. There is still less confidence "in predictions of policy changes resulting from changes in opportunity costs" because these "depended on matters like the political will of leaders and the strength of social contract about which little that is useful is known."

22. *Dollar Costing*, pp. 15, 17.
23. The titles of some of the research reports published during this period suggest the range of the effort: "O&M–General Purpose Vehicles"; *Soviet Military Aircraft Maintenance; Soviet Naval Ship Maintenance*.
24. The revision was prompted by a statement by an eminent Soviet scientist (see note 28, appendix A).
25. David Holloway, *The Soviet Union and the Arms Race*, p. 134.
26. CIA/MEAP, *Second Annual Report to the Deputy Director for Intelligence*, pp. 4–5. In its second annual report MEAP highlighted the weaknesses of OSR's RDT&E estimating methodology and recommended that CIA stop publishing such estimates until a better methodology was available. In later reports MEAP continued to urge that an alternative approach be developed and that OSR at least report the spending estimates with and without RDT&E included.
27. MEAP Chairman Ivan Selin to DCI Casey, Dec. 17, 1984, and Nov. 26, 1985.
28. CIA/OSR, *Methods for Estimating Soviet Defense Expenditures*.
29. The topics covered in the internal CIA/OSR working papers under the title *Methods for Estimating the Costs of Soviet Defense Activities* follow:

Vol. I, *Order of Battle Estimating Methods*, SR MEAC WP, *June* 1978.
Vol. II, *Manning Factor Estimating Methods*, MEAC WP, *Apr.* 1978.
Vol. II, *Manning Factor Estimating Methods*, SOVA/EAP WP, *Apr.* 1983.
Vol. 3, *Procurement Estimating Procedures*, SR MEAC WP, *May* 1977.
Vol. III, *Activity Level Estimating Methods*, SR/MEAC WP, *Aug.* 1980.
Vol. IV, *Procurement Estimating Methods*, SR/MEAC WP SR M80-10105 JX, TCS 6591/80, *Aug.* 1980.
Vol. IV, *Procurement Estimating Methods*, SOVA/EAP WP, TCS 3919/83, *Apr.* 1983.
Vol. VII, *$ Cost Estimating Methods*, SOVA/EAP WP TCS 3922/83, *Apr.* 1983.
Vol. VIII, *Ruble Cost Estimating Methods*, SR/MEAC WP, SR M80-10106C, TCS 6591/80, *Aug.* 1980.

30. CIA/MEAP, *Report of the Military-Economic Advisory Panel to Admiral Stansfield Turner, Director of Central Intelligence*. MEAP member Abraham Becker of the RAND Corporation drafted the panel's reply. He said the dollar estimates were needed as a stepping stone in deriving ruble values for much of procurement and O&M. Dollar comparisons provided a measure of the relative size of the "annual flow of inputs into the military sector." They were not precise measures of comparative additions to military effectiveness but were a "useful—if limited—measure" (appendix 2).
31. Memorandum, CIA/MEAP, "Military Economic Advisory Panel Meeting of Apr. 13–16, 1978." The Office of Soviet Analysis (SOVA) was established in late 1980.

32. MEAP Chairman Ivan Selin to DCI William Casey, memorandum, "Review of Soviet Military-Economic Analysis."
33. DDI to Selin, July 29, 1982. Gates wrote: "At my meeting with the MEAP in May, I described my concerns about the reliability and use of the analysis of Soviet defense expenditures carried out by our Office of Soviet Analysis. I indicated my intention to have a thorough review of this general area as a means to determine whether the effort should be continued in view of its controversial work and the cost of that work."
34. The findings were summarized in two reports by the Working Group on Soviet Military Economic Analysis: an overall evaluation *(Report of the Working Group on Soviet Military Economic Analysis,* July 20, 1983) and a report on methodology *(Report of the Methodology Panel of the Working Group on Soviet Military Economic Analysis,* July 1983).
35. *Report of the Working Group,* pp. 3–8.
36. *Report of the Methodology Panel,* pp. 8–14.
37. "If procurement costs go up because of falling productivity, bottlenecks, technical or production problems, or other such problems, the current methodology will not catch these increases until the Agency succeeds in transferring the estimates to a new price base." Ibid., p. 12.
38. "Since the military economic estimates are component analyses, they have suffered badly from the redirection of effort under the SOVA reorganization and the consequent reduction in the number of analysts doing these estimates." Ibid., p. 14.
39. *Report of the Soviet Working Group,* p. 1.
40. Ibid., pp. 21–24. See chapter 5 for CIA's estimate of the cost of empire.
41. *An Evaluation of CIA Work on Soviet Economic Capabilities, Problems, and Prospects,* p. 101. The PFIAB-sponsored study assessed CIA's reports on Soviet economic prospects as being too negative, whereas the Rowen panel two years later found them overly optimistic. The PFIAB panel found fault not in CIA's measurement of Soviet economic growth but in its discussion of its social and political implications. The Rowen panel found that CIA's measures of Soviet economic growth might be too high but, like the PFIAB panel, found "little basis for inferring heightened expressions of political discontent resulting from even poorer economic performance."
42. CIA/OSR, Programs Analysis Division. "Monthly Activity Report for March 1976," Apr. 26, 1976.
43. CIA/OSR, Memorandum for Director, National Foreign Assessment Center, "OSR's Analytical Resources Assignments by Topic, Category, and Country."
44. Testimony before the Selin panel, Nov. 20, 1982, pp. 27, 29–30.
45. Testimony before the Selin panel, Dec. 10, 1982, pp. 6–8.
46. CIA/DI, *A Comparison of Soviet and U.S. Defense Activities, 1972–1981,* p. 4.
47. Interviews with Steven Altobelli, May 12, 1995, and J. Michael Martin,

June 1995. News of the existence of the new estimate was soon circulating in the intelligence community, however. Paul Berenson, a Pentagon official, told the Becker panel (Dec. 20, 1982) that he had learned within the past few weeks that Soviet defense procurement had been flat for five years. If accurate, he said, the new estimates were an example of how the CIA estimates "are very important and boy, that's a totally different picture." According to Donald Burton (testimony before the Selin panel, Mar. 24, 1983), when CIA revised its estimates it left Berensen and Marshall, director of net assessments in DoD, "up in the air" because they were preparing to use the estimates in the budget debate.

48. CIA/DI, *Soviet Defense Spending: Future Trends and Recent Prospects.*
49. Interview with Rae Huffstutler, June 3, 1994.
50. Interview with J. Michael Martin, June 1995. The assistant national intelligence officer for general purpose forces was especially skeptical, Martin reported. "They produce a lot of tanks, don't they? Tanks are expensive, aren't they?"
51. Interview with Terry Dunn, Apr. 1996.
52. Memorandum to Chairman, NIC, Mar. 9, 1983. Cost overruns on the ruble side, of course, would not have affected the dollar estimates, which also portrayed a slower growth in the *size* of Soviet defense programs.
53. Interview with Robert Gates, May 12, 1994. Gates noted that this provided additional proof that CIA was willing to deliver bad news.
54. Herbert Meyer, Memorandum for Director of Central Intelligence, Deputy Director of Central Intelligence and Chairman, National Intelligence Council, "Fred Ikle's Concern about Intelligence Community Soviet Cost Estimates."
55. Nonetheless considerable doubts that the slowdown was real remained within the agency, and strenuous denials were heard outside. James Steiner, then a SOVA branch chief, argued that something might have happened to the linkage between output and input in the defense industry in the late 1970s. In other words procurement may have been flat in terms of output but not in terms of the resources consumed in defense production. Testimony before the Becker panel, Feb. 17, 1983, pp. 3, 11. Derk Swain, the senior analyst in charge of SOVA's presentations to the panel, wondered whether, given that a change of four lines—the major procurement items—would remove the slowdown, the observed changes were enough to support or deny its existence. Testimony before the Becker panel, Mar. 17, 1983, p. 61.

William Lee (in "The Real Implications of the CIA/DIA Joint Statement to the JEC," May, pp. 5ff) asserted that his estimate of Soviet defense spending showed continued rapid growth after 1975. Although the number of weapons produced may have leveled off or even declined, procurement costs had been driven up by the trend toward increased complexity and advanced technology embodied in new weapons. Patrick Parker and Rosefielde declared that "careful analysis" indicates that "the

Agency's new estimates are wrong" and that military superiority continues to be the USSR's objective. They listed ten reasons for rejecting CIA's estimates, including the intrinsic difficulties of the building-block approach, the instances of past changes in CIA estimates, agency underestimates of the pace of product improvement in Soviet weaponry, and a supposed inconsistency between CIA estimates and Soviet statistics on output and employment in the machine-building industry. Parker and Rosefielde, "Soviet Arms Procurement Strategy in the Eighties: Conflicting Perceptions of the Soviet Arms Buildup," *Russia* 12 (1986): 49–50.

56. Robert M. Gates, "Soviet Defense Costing Effort."
57. CIA/DI, *Soviet Defense Spending—Recent Trends and Future Prospects*, p. iv.
58. The authors further stated that unanticipated problems (manufacturing and supply difficulties) could not explain the slowdown because it had gone on so long. They also pointed out that a plateau in procurement (the flow of weapons into military stocks) did not prevent stocks of weapons and equipment from growing substantially.
59. Raymond L. Garthoff, *Détente and Confrontation*, p. 596.
60. Raymond L. Garthoff, *The Great Transition*, p. 41.
61. Orrin G. Hatch to Hon. William J. Casey, Mar. 28, 1984. Senators Jesse Helms, Strom Thurmond, Roger Jepsen, John East, and Steve Symms were the other signatories. The new estimate, of course, did not represent a cut in Soviet defense spending but a slower rate of growth. Casey wrote back saying that despite the reduced growth, the dollar costs of Soviet defense activities were still far higher than those of the United States. In any case, he said, one should look at outlays over several years and the accumulated stocks of Soviet weapons and equipment. Casey to Hon. Orrin G. Hatch, Apr. 18, 1984.
62. CIA/SOVA, Memorandum for Deputy Director of Intelligence, "Record of Published Estimates of Soviet Defense Spending."
63. Ibid.
64. CIA/NFAC, *The Development of Soviet Military Power: Trends since 1965 and Prospects for the 1980s*, pp. 37, 95.
65. Interviews with J. Michael Martin (June 1995) and Terence Dunn (May 3, 1995).
66. Ibid.
67. The methodology is set out in CIA/SOVA, Memorandum for the Record, "Estimating Soviet Defense Spending in Current Rubles."
68. Intelligence focus on projections during the McNamara years at the Pentagon, which resulted in intelligence products like the National Intelligence Projections for Planning, provided an important exception to this generalization.
69. CIA/OSR, internal memorandum, "A Reduction in Soviet Military Production?" The memorandum maintained that if the projections were

correct, the possible explanations required investigation: "a recognition by the Soviets that they have 'enough'; a transfer of resources from military production to RDT&E; a desire to remedy civilian economic problems at the expense of the military."

70. Interview with James Barry, Oct. 13, 1995.
71. CIA/OSR, "OSR's 1977 Projection." Undated briefing paper.
72. The briefing paper noted, for example, that projections of fighter and bomber aircraft made in 1971 implied a near shutdown of four major airframe plants—an event highly inconsistent with observed Soviet practice.
73. Interview with Derk Swain, Nov. 15, 1995. Only spending was increased. No changes were made in the underlying forces and programs.
74. CIA/DI, *Projecting Soviet Military Forces and Weapons Procurement*.
75. Each judgment was informed by all-source intelligence on the projected system. Often different levels of confidence were attributed to the different aspects of the projection. For example, the analysts might have high confidence that a given weapon might be produced for at least five years but less confidence that its production run would last for eight years. The analyst was asked to use four categories of uncertainty: A-100 percent certainty, B-75 percent, C-50 percent, and D-25 percent. CIA/DI, *Projecting Soviety Military Forces*, p. i.
76. Ibid., p. 11.
77. Ibid., p. 12.
78. Ibid., p. 4. Within each aggregate (for example, spending on naval forces), a number of combinations of production programs were possible. The paper explained, "In some respects, our method is analogous to projecting the outcome of a series of coin tosses. There is a sound statistical basis for saying we have a 'best estimate' of 50 heads in 100 tosses. We have no basis, however, for predicting the outcome of each individual toss, other than to say there is a 50 percent chance of heads."
79. CIA/DI, "The Cost of Projected Soviet Strategic Forces," p. 1.
80. Ibid., pp. 15–22.
81. Interview with Steven Altobelli, May 12, 1995.
82. CIA/SOVA, Memorandum for DDI from Douglas J. MacEachin, Director of Soviet Analysis, "NIE 11-3/8: Force Projections." The memorandum relied heavily on the analysis that had been carried out in preparing SOVA's 1986 force projection paper. In *From the Shadows* Gates, citing this memorandum, comments, "The lack of communication between the economists and the military experts seemed hopeless" (p. 386). He implies the problem was in SOVA, whereas the disagreement was between SOVA's analysts and the national intelligence officers responsible for the military estimates and the DoD representatives assigned to the military NIEs.
83. NIE 11-3/8-86 II, *Soviet Forces and Capabilities for Strategic Nuclear Conflict through the Mid-1990s*, pp. iv–10.

84. Ibid., pp. iv–12.
85. Testimony before the Selin panel, Mar. 24, 1983.
86. The first edition of *Soviet Military Power* appeared in Sept. 1981.
87. Testimony before the Selin panel, Feb. 24, 1983, p. 13.
88. Carl G. Jacobsen, "Soviet Military Expenditure and the Soviet Defense Burden," in *World Armaments and Disarmament, SIPRI Yearbook 1986*, p. 264.
89. Franklyn D. Holzman, "How CIA Concocts Soviet Defense Numbers," *New York Times*, Oct. 25, 1989.
90. Steven Rosefielde, *False Science: Underestimating the Soviet Arms Buildup*, p. 39.
91. SOVA replied: "The comparison between U.S. and Soviet defense expenditures is one of our most frequently requested products by DoD and the Congress. It is controversial, the answers are occasionally misused, and it may inspire as much public confusion as insight. But the dollar cost estimates are still the most commonly used way of aggregating the disparate parts of the Soviet military in a manner that allows comparisons with the defense activities of the US and other countries. They have proven consistent and useful as a measure of the trends in Soviet defense activity over time. They are also the only means available to assess Soviet military programs on a mission or resource basis." CIA/SOVA, Memorandum for Director of Central Intelligence, Deputy Director of Central Intelligence, "Dollar Cost Analysis of Soviet Defense Estimates."
92. CIA/SOVA, Memorandum for Chief, Econometric Analysis Division, "Chronology of Events on the Subject of Soviet/U.S. Defense Comparisons."
93. Ibid.
94. Interview with Robert Gates, May 12, 1994.
95. Robert M. Gates, "Statement of Robert Gates, Deputy Director for Intelligence, Central Intelligence Agency," in *Allocation of Resources in the Soviet Union and China, 1984*, pp. 19–20.
96. CIA/SOVA, "Chronology of Events."
97. Interview with Robert Gates, May 12, 1994. In his book Gates declared that although "as a noneconomist, nonstatistician" he found it hard to quarrel with the agency's methodology, he thought that "in this communist variant of Sparta," the share of defense in Soviet GNP was between 25 and 40 percent (p. 318).
98. For example John Helgerson, the DDI in 1989–93, said he was "never particularly comfortable" with the work. He felt best about the physical estimates, less good about the ruble estimates, and doubtful about the dollar estimates, which he called "house of cards analysis." Interview, Apr. 15, 1994. Attacks on the agency's estimates of the dollar equivalent of Soviet GNP as overstating its size introduced considerable confusion. Thus Gates maintains that CIA "overstated the size of the Soviet economy and relatedly underestimated the burden of military expendi-

tures on the economy and society" (*From the Shadows*, p. 564). As we have argued, size and burden are separate questions. Burden can only be assessed in ruble prices, and CIA estimates of ruble GNP agree well with alternative estimates by western and Russian statistical authorities. If anything these estimates are understated because they fail to take full account of second economy activities and the contributions of four million employees who did not appear in Soviet employment statistics until 1990. Timothy Heleniak, "Puzzling Soviet Labor Force Statistics: Declining State Sector Employment and Employment in the Ministry of Defense," p. 1.

99. "Overall . . . the CIA does an excellent job of estimating Soviet military expenditures." *Report of the Working Group*, p. 1.

100. William J. Casey to Ivan Selin, Nov. 29, 1983. Casey said DDI Gates had told him the report had received wide distribution, including all members of PFIAB, the chairman of the Senate Select Committee on Intelligence, and the House Permanent Select Committee on Intelligence. Casey did not say whether the DoD or the NSC had been on the distribution list.

101. CIA/DDI, Note to Director of Central Intelligence, Deputy Director of Central Intelligence, "Soviet Procurement/Ruble Costing Paper."

102. Ibid.

103. James Locher, testimony before Selin panel, Feb. 24, 1983, p. 8.

104. George Wilson, "Intelligence Units Agree They Disagree," *Washington Post*, June 14, 1984.

105. *Allocation of Resources in the Soviet Union and China–1984*, pp. 18, 237–39.

106. Note from DDI Gates to director of SOVA, May 31, 1983. SOVA replied that DIA's figures showing high rates of growth for procurement in 1976–81 were based on faulty production numbers. When corrected, SOVA said the production numbers generated the same trend in costs as SOVA's estimate. CIA/SOVA, Memorandum for Deputy Director for Intelligence, "Soviet Defense Costs."

107. *Allocation of Resources in the Soviet Union and China–1985*, p. 102.

108. Holzman, "Politics and Guesswork," pp. 118, 124–26.

109. CIA/DDI, Draft Memorandum for the Record, "Secretary of Defense/DCI Discussion of Soviet Defense Expenditures, 20 July 1984." The final version of the memorandum omitted these references to the declining utility of the comparisons to DoD. A DIA manager told agency analysts that DIA's early 1984 announcement of a recent rise in the growth of Soviet procurement had been prepared, declassified, and released "at the request of the Secretary of Defense." Michaud said he thought DIA had been blindsided by CIA's 1983 slowdown paper: "We (CIA) simply had no idea of the problems our estimate had caused DIA." CIA/SOVA, Memorandum for Director of Soviet Analysis, "Meeting with Norb Michaud, Chief, DB-4E, on Terms of an Agreement on Defense Cost Comparisons."

110. Richard Kaufman, the JEC staff representative, observed that the decision to withhold the defense comparisons appeared to be motivated by the fact

that U.S. and Soviet defense budgets were now converging. CIA/SOVA, Memorandum for the Record, "Meeting with Richard Kaufman, Associate Staff Director, Joint Economic Committee."

111. CIA/SOVA, Memorandum for Deputy Director for Intelligence, "The Release of Dollar Cost Comparisons of U.S. and Soviet Defense Costs." Ivan Selin, in his capacity as MEAP chair, also warned Casey about the consequences of cutting off the dollar estimates. "On a practical level," he said, if CIA gives comparison piecemeal to qualified consumers, they will be used without the caveats included in CIA publications. The dollar comparisons are part of a large analytic framework and "to delete one or another of these components is to damage the integrity and utility of the estimate package as a whole." Ivan Selin to DCI Casey, Dec. 17, 1984.
112. DIA, Memorandum for the Deputy Director for Intelligence, CIA, "Dollar Estimates."
113. CIA/SOVA, Memorandum for Deputy Director for Intelligence, "Recommended Response to Ikle Request."
114. Gates to MacFarlane, Feb. 15, 1985.
115. CIA/SOVA, Memorandum for Deputy Director for Intelligence, "Weinberger Briefing on Soviet Defense Spending Issues."
116. CIA/DDI, Note to Director, SOVA, "Soviet Defense Expenditures."
117. In 1986 the ratio of USSR to U.S. costs was 1.05 to 1.07 under the expanded definition and 1.04 under the traditional definition. CIA/DI, *A Comparison of Soviet and U.S. Defense Activities, 1973–1987*, p. 26.
118. CIA/DI, *A Comparison of Warsaw Pact and NATO Defense Activities, 1976–1986*.
119. CIA/DDI, "Note to Director of Soviet Analysis," Dec. 10, 1987.
120. CIA/SOVA, Memorandum for Deputy Director for Intelligence, "Request for Warsaw Pact-NATO Defense Comparisons."
121. CIA/DDCI, Memorandum for Deputy Director for Intelligence, "Comparison Warsaw Pact and NATO Defense Activities, 1976–1986," Nov. 5, 1987.
122. CIA/DI, *Soviet Economic Problems and Prospects*. The paper was an unclassified version of a study commissioned by the NIC.
123. Ibid., pp. ix–x.
124. CIA/NFAC, *Development of Soviet Military Power*.
125. Ibid., p. v. The baseline projection called for growth in real defense outlays of 2–4 percent per year through the 1980s.
126. CIA/DI, *Soviet Defense Spending*, p. v.
127. See "Charges to Review Panels," above.
128. The results of the redrafting appeared in Henry S. Rowen, "Prepared Statement of Hon. Henry Rowen," in *Allocation of Resources in the Soviet Union and China, 1982*, pp. 192–205.
129. CIA/DDI, Note to Paul Ericson, "Balancing Soviet Resources in 1986–1990." Gates was struck by the upward tick in growth under Andropov, which was partly due to Andropov's discipline campaign, and believed

that Gorbachev's increased pressure on managers and workers to perform could raise the economy's potential in both the short and long run.

130. Robert M. Gates and Lawrence K. Gershwin, "Soviet Strategic Force Developments." Testimony before a Joint Session of the Subcommittee on Strategic and Theater Nuclear Forces of the Senate Armed Services Committee and the Defense Subcommittee of the Senate Committee on Appropriations, June 26, 1985.

131. The director of SOVA summarized his office's views at this point in a briefing paper. To begin with, Gorbachev was determined to improve the USSR's production base. This would require massive investment, which taken together with other economic requirements "appeared to leave little or no room for increasing defense spending procurement above its recent rates." This strategy (relying on investment to raise productivity) was at least moderately and quite likely highly risky because it would "require the Soviet system to do what it has never done well." Douglas MacEachin, "Gorbachev's Economic Strategy and Soviet Defense Spending."

132. CIA/DI, *Gorbachev's Modernization Program: Implications for Defense*.

133. Ibid., pp. 9–10. There was evidence that Soviet leaders expected Gorbachev's plans to affect defense. MacEachin's Dec. 1985 briefing notes reported: "Since May, a number of Soviet officials have said that Gorbachev's industrial modernization program will impinge on defense. Some of these statements clearly were intended to be conveyed to U.S. and other Western officials, evidently as presummit tactics designed to generate pressure for the U.S. to 'capitalize on the opportunity.' Others, however, were obtained from controlled, clandestine sources who were reporting their understanding of the thrust of high level economic policy discussions in the USSR."

134. Interviews with Douglas MacEachin and Derk Swain, Dec. 11, 1995. The briefing for Shultz also included a discussion of SOVA's new force projection methodology, which appeared to be of great interest to the secretary.

135. Ibid.

136. CIA/SOVA, "The State of the Soviet Economy." The director of the Office of Soviet Analysis, in transmitting the memorandum to Gates, wrote: "attached is a first cut at the memo you asked for on the Soviet economy. I hope it at least captures the flavor you were looking for. The figures are all consistent with what we have been publishing; this memo simply takes the 'half full' rather than the 'half empty' approach."

137. CIA/SOVA, Memorandum for Director of Soviet Analysis, "CIA's Dollar Cost Work: Where Do We Go from Here?"

138. CIA/SOVA, "Talking Points for the DDI on Curtailing Our Work on Estimating Dollar Cost Comparisons of U.S.-Soviet Defense Activities."

139. Andrew Marshall to DDI John Helgerson, June 25, 1991. Marshall, in fact, asked that the comparisons be extended to other countries.

140. John Helgerson to Andrew Marshall, July 9, 1991. Helgerson's stated

reasons were peculiar: first, the estimates were an artificial construct that had no relationship to the burden of defense (they were never intended as such), and second, they are frequently misinterpreted or misused as an "ostensible measure" of defense burden and a reflection of military capability.

## Chapter 5. The Spending Estimates

1. The error could be reduced if the machine-readable files on detailed order-of-battle and military production for the pre-1965 period had not been destroyed. What remains are reliable cost estimates (the "historical best") at a higher level of aggregation. Thus a shift in the mix of procurement of aircraft or ships and boats would alter the average ruble-to-dollar ratios for aircraft or ships and boats to the extent that the ratios for individual aircraft or ships differed.
2. RDT&E is excluded because the methodology supporting the estimate changed drastically in the mid-1980s.
3. As explained above, the broad definition of defense includes the activities that would be included under U.S. defense spending, as well as the costs associated with civil defense; railroad, construction, and MVD troops; and spending on space activities that in the United States would be funded by NASA.
4. Stephen M. Meyer, "Economic Constraints in Soviet Military Decision-Making," in Henry S. Rowen and Charles Wolf, Jr., eds., *The Impoverished Superpower*, pp. 203–206.
5. The definitions of the missions shifted over time. Frequently spending on general purpose forces (ground forces, nonstrategic navy, and tactical aviation) was broken down into its elements.
6. The difference is actually larger than $65 billion because it proved impossible to add the weapons component of Atomic Energy Commission expenditures (in 1972 prices) to the federal outlays for national defense as reported in the publications of the Department of Commerce's Bureau of Economic Analysis.
7. With growing interest in power projection capabilities in the 1970s and 1980s, the estimates identified a mobility forces element of the general purpose forces mission. It included airlift and sealift activities and military port operations.
8. CIA/OSR, Memorandum for Director, Strategic Research, "NATO–Warsaw Pact Defense Cost Comparisons."
9. *Report of the Working Group*, p. 7.
10. Ibid., pp. 7–18.
11. John J. Maresca, Memorandum for Director, Office of Soviet Analysis, CIA, and Director, Office of European Analysis, CIA, "Study on NATO–Warsaw Pact Defense Activities, 1976–1985."
12. CIA/DDCI, "Comparison Warsaw Pact and NATO Defense Activities."

13. CIA/SOVA, Memorandum for Deputy Director for Intelligence, "Our Comparison of NATO and Warsaw Pact Defense Activities."
14. We exclude RDT&E because it is the least reliable component of the estimate and one that is most likely to mask changes in trend.
15. Harry Gelman, in *The Brezhnev Politburo and the Decline of Détente* (p. 96ff), cited three crises in 1967, 1968, and 1969 that reminded the leadership of "the transcendent importance of the military priority in Soviet resource allocation." The three were Yegorychev's 1967 challenge to Brezhnev claiming his defense policies were too weak, the Czech invasion, and the firefights with the Chinese on Damanskiy Island.
16. Holloway, *Soviet Union and Arms Race*, p. 35.
17. Gelman, *The Brezhnev Polituro*, pp. 46–48.
18. It could also have signified a recognition that a continued rise in the share of investment in GNP was self-defeating—that returns to capital were falling faster than capital stock could be increased.
19. Because of a declining pool of draft-eligible males, the task of conscripting enough young people for the Soviet military forces was already a problem, as a series of CIA papers argued.
20. Gelman, *The Brezhnev Politburo*, pp. 176–81.
21. For example, Nikolay Ogarkov, chief of the General Staff, wrote in *Kommunist* (July 1981) that the danger of war was real, not merely possible.
22. Michael McGwire, *Perestroika and National Security*, p. 318. Ben Fischer chronicles the development of the Soviet war scare through 1983, when he concludes that "a full-scale war scare unfolded in the USSR" following the shoot-down of the KAL Airliner in September. *A Cold War Conundrum*, p. 27.
23. Ibid, p. 75. MccGwire writes that the Soviets saw SDI as a way of restoring U.S. military superiority.
24. Ibid., pp. 116–17.
25. Ibid., pp. 41–42, 118ff.
26. Gorbachev, *Pravda*, Dec. 10, 1989.
27. Abraham Becker, in *The Burden of Soviet Defense: A Political-Economic Essay*, put the issue succinctly:

> *Opportunity costing implies efficient operation—at a point on the society's production frontier, the focus of real production possibilities with given resources and technology. At that point, the relative prices of any pair of goods and services produced correspond to the rate at which one of those goods may be transformed into the other at the margin by relocating existing factors of production. Factor relative prices in turn correspond to the ratios of the value of their marginal products. In this idealized context, valuation of resources at marginal cost provides a true measure of the economy's production potential, its ability to produce varying mixes of goods*

and services given its current resource endowment. Here marginal costs are opportunity costs, for relative product prices are also measures of the rate of transformation into alternative uses. By the same token, the value of military expenditure will also equal the value of civilian output foregone. (p. 6.)

28. Marshall had long believed that even CIA's broad definition of defense activity did not represent the full extent of the burden of Soviet national security programs. See Andrew W. Marshall, "Commentary," in *Gorbachev's Economic Plans*, vol. 1, *Study Papers*, pp. 481–84; and his testimony before the Becker panel. His interest helped spur other research on the question. See David F. Epstein, "The Economic Cost of Soviet Security and Empire," in Rowen and Wolf, Jr., *The Impoverished Superpower*; Charles Wolf, Jr., et al., *The Costs and Benefits of the Soviet Empire*, 1983; Charles Wolf, Jr., et al., *The Costs and Benefits of the Soviet Empire*, 1986; and Charles Wolf, Jr., "The Costs and Benefits of the Soviet Empire," in Rowen and Wolf, Jr., *Future of the Soviet Empire*.
29. CIA/DDI, Note to Director, SOVA, Apr. 1, 1985, "Soviet Defense Expenditures." DDI Gates wrote the Director of Soviet Analysis that the next dollar-comparison paper should contain a section on a "broader view of the Soviet burden"—one that would consider "the cost of maintaining the empire (including economic subsidies to Cuba, Vietnam, Ethiopia, Nicaragua and so on), and factoring in a relevant portion of the cost for aircraft, the Soviet merchant marine and so on with respect to their military utility and use as a reserve force." Two weeks later DCI Casey wrote Secretary of Defense Caspar Weinberger that the next costing paper would take a new look at the Soviet defense burden and that "Our guess is that this kind of alternative will produce a Soviet defense burden that is on the order of 20 percent or more of GNP rather than the 13–16 percent we and DIA have been carrying." Executive Registry, 85-1588/2, Apr. 16, 1985.
30. CIA/DI, *A Comparison of Soviet and U.S. Defense Activities, 1976–1985*, pp. 20–22, 37–38.
31. CIA/DI, *Defense's Claim on Soviet Resources*, pp. 2–3, 11–13.
32. For example, a 1968 CIA memorandum reported: "Although military spending overall showed only modest growth under Khrushchev, a heavy drain was imposed on those resources urgently needed for economic growth. Rising outlays on advanced hardware and for R&D drained resources from the industrial investment program and hampered Khrushchev's efforts to improve technology in civilian industry. . . . This in turn, intensified pressures to revamp the Soviet planning structure and the way the economy was managed." CIA/DI, *The Price of Strength: Broader Soviet Force Goals Driving Up Defense Spending*, pp. 10–11.

Later the agency said that even though the share of defense in GNP was declining, some Soviet leaders "could be even more concerned with defense costs now than they have been in the past." Again, the technological drain was featured in the discussion, but the paper went on to talk about the hidden costs to the civilian economy of giving defense priority in the industrial supply system and "a new factor"—the commitment of the Brezhnev regime to improving the lot of the Soviet population. CIA/DI, *Soviet Spending for Defense: An Annual Review*, vol. 1, *Trends in Ruble Expenditures*, pp. 28ff.

33. Some of the CIA papers that modeled the impact of defense spending on economic growth included CIA, *The Economic Impact of Soviet Military Spending*; CIA/NFAC, *Simulations of Soviet Growth Options to 1985*; CIA/DI, *USSR: Economic Projections, 1982–1990*; and CIA/DI, *USSR: Economic Projections through 1990—A New Look*.

34. Rush V. Greenslade, "The Many Burdens of Defense in the Soviet Union," *USSR Monthly Review* (Apr. 1982): 57–63.

35. Ibid, 58. Greenslade maintained that "in an economy in a chronic state of 'disequilibrium' *opportunity cost* has several different values, depending on which alternative goods are valued, and it is quite uncertain which of the many *opportunity costs*, if any, is measured by *resource cost*."

36. Russia's experience with defense conversion in the post–Cold War period lends some weight to Greenslade's argument. Managers of defense-industrial enterprises proved to be poorly equipped to operate outside a preferential environment, and their plant and equipment was often unsuited for civilian production.

37. NIE-11-4-54, *Soviet Capabilities*, p. 9.

38. NIE 11-4-63, *Soviet Capabilities and Policies, 1962–1967*, pp. 25–27.

39. NIE 11-4-65, *Main Trends in Soviet Military Policy*, p. 10.

40. The NIEs devoted to strategic forces were especially dismissive of the possibility that economic constraints could affect their forces. Thus NIE 11-3/8-76, *Soviet Forces for Intercontinental Conflict through the Mid-1980s*, vol. 1, *Key Judgments and Summary*, said, "We see no evidence that economic considerations would inhibit the expansion of strategic programs" (p. 21). Again, in 11-3/8-83, *Soviet Capabilities for Strategic Nuclear Conflict, 1983–93:* "Nor do we believe that domestic economic difficulties will bear significantly on the size and composition of future Soviet strategic forces because of the high priority the Soviets place in such forces" (p. 26).

41. CIA/OER, Memorandum for the Record, "Response to Questions from the Proxmire Briefing."

42. Douglas MacEachin, "Gorbachev's Economic Strategy and Soviet Defense Spending," Memorandum.

43. CIA/SOVA, "Overview: The Impact of the New Five-Year Plan on the Defense Sector."

44. Topical interview with Marshal Sergey Fedorovich Akhromeyev. Moscow television service in Russian, 1700 GMT, Oct. 9, 1989.

## Chapter 6. The Critics

1. *Report of the Working Group; Report of the Methodology Panel of the Working Group on Military Economic Analysis;* Berkowitz, et. al, "An Evaluation of the CIA's Analysis of Soviet Economic Performance, 1970–1990," *Comparative Economic Studies* 35, 2 (Summer 1993): 33-48.
2. Testimony before the Selin panel, Jan. 13, 1983, pp. 10–11.
3. Testimony before the Becker panel, Dec. 20, 1982, p. 15.
4. *Report of the Working Group,* p. 6. In his testimony before the working group's consumer panel (the Selin panel), Helmut Sonnenfeldt of the Brookings Institution said that besides the general interest in what the Soviet defense effort meant to the economy, "a decisive reason (for interest in the subject) is what this does to the whole Soviet body politic and what it may do to choices that Soviet political leaders make or have to make and consequently what possible conceivable openings this question of economic aspects of defense may provide for American policy to influence Soviet behavior and developments." Testimony before the Selin panel, Jan. 13, 1983, p. 1.
5. *Pravda,* June 11, 1989, p. 5.
6. See "Defense in Industrial Statistics" in this chapter.
7. See Abram Bergson, *The Real National Income of Soviet Russia since 1928,* pp. 103–26.
8. Testimony before the Becker panel, Dec. 20, 1982, p. 15.
9. Igor Birman, "Statement to the Working Group of the CIA on Soviet Military Expenses," p. 3.
10. CIA/DI, *Soviet Spending for Defense: An Annual Review,* vol. 2, *A Monetary Comparison of Soviet and U.S. Defense Activity,* p. 9.
11. *Comparison of Soviet and U.S. Defense Activities, 1972–1981,* pp. 6–7.
12. Secretary Brown noted that the dollar and ruble measures of Soviet programs complemented physical measures of the U.S.-Soviet military balance with "an understanding of such qualitative factors as military doctrine, training practices, strategy and deployed technology."
13. Marshall to John Helgerson, June 25, 1991. In his appearance before the Becker panel, Firth emphasized that the unique contribution of the dollar and ruble estimates was that they were the only way of aggregating the various components of defense activity and avoided the tendency in intelligence to focus only on things that are going up. Testimony before the Becker panel, Dec. 10, 1982, p. 2.
14. Testimony before the Selin panel, Jan. 13, 1983, p. 12; Jan. 27, 1983, p. 2. Testimony before the Becker panel, Dec. 20, 1982, p. 11.
15. *Report of the Working Group,* p. 6.
16. William T. Lee, *CIA Estimates of Soviet Military Expenditures—Errors and Waste,* pp. 70, 72. Indeed, he thought the additional activities included in the broad definition were a CIA "fudge factor." Nonetheless he cited a number of independent Soviet estimates of the USSR's defense outlays as support for his own estimates even though these independent Soviet

estimates generally included the activities in CIA's broad definition and more.
17. Henry S. Rowen, "Biting the Bullet and Other Hard Choices for Moscow," p. 15.
18. Igor Birman, "The Size of Soviet Military Expenditures: A Methodological Aspect," p. 12.
19. Testimony before the Becker panel, Dec. 10, 1982, p. 2.
20. Testimony before the Becker panel, Nov. 20, 1982, p. 19.
21. Rowen, "Biting the Bullet," p. 9.
22. Ibid., p. 16.
23. Parker and Rosefielde, "Soviet Arms Strategy," p. 55.
24. Testimony before the Selin panel, Mar. 3, 1983, pp. 11–20. See also William Odom, "The Riddle of Soviet Military Spending," *Russia* 3 (1981): 53–58.
25. See, for example, Holzman, "Are the Soviets Really Outspending," pp. 87–90; and Holzman, "Soviet Military Spending: Assessing the Numbers Game," *International Security* 6, 4 (Spring 1982): 78–93.
26. A reviewer of this book, Abraham Becker, points out that if subsidies were important in defense industry and especially high for more complex weapons and equipment, the ruble value of U.S. defense spending in factor cost rubles would be higher than shown in CIA estimates and the rate of growth of U.S. programs in rubles would also probably be higher.
27. The other resource categories included in defense spending did not present the same problems as procurement (except for the spare parts in O&M). Enough was known about military pay rates and price indexes for fuel and power, transportation, construction, and the wages of civilian employees and RDT&E workers to move from a 1970 to a 1982 price base.
28. Those included prices for aircraft (25), helicopters (8), surface-to-air missiles (15), air-to-air and air-to-surface missiles (7), ballistic missiles (7), land arms (28), electronics (44), and ships and submarines (8).
29. Estimating product group ruble-to-dollar ratios:

$$\frac{Ry}{\$} = \frac{Rb}{\$} (NW)^{(p-b)} (SW)^{(y-p)} (TF)^{(f-b)}$$

$b$ = year of price base

Set of observations

$Ry$ = ruble price of weapons system in year y

$ = price of weapon system in constant dollars

$y$ = year price pertains to

$p$ = year first permanent price was set

$f$ = year of first flight or test of weapon system (a proxy for technological complexity)

Parameters to be estimated for each product group:

$NW$ = inflation in new weapon system prices (i.e., $NW$ = 1.04 means 4 percent inflation)

$SW$ = inflation in series production weapons prices

$TF$ = technological factor
$Rb$ = ruble price of weapon system in year b

30. The inflation in 1971–80 indicated by the regression analysis was as follows:

|  | Average Annual Percentage Increase in Prices for Systems in Production | Percent Increase in Price When New System Is Introduced |
|---|---|---|
| Aircraft | 2.5 | 7.2 |
| Helicopters | 16.6 | 14.1 |
| SAMs | 11.0 | 20.2 |
| Electronics | -10.7 | 25.7 |
| Land arms | 0.7 | 0.4 |

31. A. I. Pozharov, "Defense Sufficiency: Military-Economic Aspect," *Military Thought-Voyennaya mysl'* 7 (1993): 34. According to Pozharov, it had "especially in recent years" tried "to impose on the Ministry of Defense those products whose manufacture was adjusted without regard for the military's requirements." A retired naval captain complained bitterly in 1992 that the MIC continued to supply weapons and equipment that were of low quality and not needed. He asserted that "this has been possible because the military-industrial mafia itself orders the weapons, equipment, and armaments, builds them itself, gets them up to the necessary level itself, carries out the acceptance itself, and itself investigates all accidents involving equipment, weapons, and armaments, including the death of personnel." Moscow Russian Television Network in Russian, 1500 GMT, July 2, 1992.

Former high Soviet defense officials have asserted that the power of the defense industrialists had long distorted military decisions. V. S. Shabanov, the deputy minister of defense for armaments, after citing some examples of world-class Soviet weapons (the MIG-29 and SU-27 aircraft), said that the situation regarding procurement could not be considered satisfactory. Equipment reliability was inadequate, to a large extent because of the orientation in the 1960s and 1970s toward "quantitative growth" of basic weapons—tanks, aircraft, and ships. V. S. Shabanov, "Noviye podkhody (New Approach)," *Krasnaya zvezda,* Aug. 18, 1989, p. 2. Gen. Viktor Kulikov, chief of the Soviet General Staff from 1971 to 1977, and Gen. Makhmet Gareyev attributed the clout of the Soviet military-industrial complex to its ability "to gradually place its own people on the staff of the Central Committee, the Council of Ministers, Gosplan and other organs, and they defended the interests of their 'native' sectors of industry in isolation from the interests of the matter as a whole, sometimes without taking into account the General Staff's proposals." Kulikov and Gareyev, "Some Issues of Military Technical Policy," *Vooruzheniye, politika, konversiya,* no. 2 (1993): 29–33.

32. The breakdown of prices follows: aircraft (124), SAMs (67), other missiles (49), radar (11), ships (13), space equipment (3), tanks (57), other land arms and vehicles (55), and small arms (7).
33. CIA/SOVA, Memorandum for the Record, "Review of Ruble Procurement Prices."
34. Generally, within a specific time period the prices of products whose output is increasing most rapidly will fall or rise more slowly than the prices of more established products. Thus a 1970 price base will give greater weight to fast-growing items than a 1982 price base.
35. Holzman, "Politics and Guesswork," pp. 109–16.
36. Ibid., p. 111.
37. Berkowitz, et al., "Evaluation of the CIA's Analysis of Soviet Economic Performance," p. 46. "If the revised base prices do in fact measure true factor cost, then the burden of defense should decline in normal expectation."
38. Lee, *CIA Estimates of Soviet Military Expenditures,* pp. 25–26, 46, 70.
39. See "Defense in Industrial Statistics" in this chapter.
40. See CIA/DI, *Guide to Monetary Measures,* p. 8.
41. Testimony before the Becker panel, Nov. 20, 1982, p. 21.
42. Testimony before the Becker panel, Feb. 18, 1983, pp. 5–6, 70–75. Rosefielde elaborated on his criticism in a 1983 paper, "The Strong Separability of Defense and Civilian Growth Potential: A Reassessment of Soviet Growth Prospects in the 1980s."
43. *Report of the Working Group,* p. 9. In addition, costs reflecting learning were preferred measures because the price indexes used to deflate U.S. defense spending in current prices left the effects of learning in the U.S. constant price series.
44. Earlier Swain had told the Becker panel that the difference would be very slight. Rosefielde testimony before the Becker panel, Feb. 18, 1983, p. 85.
45. CIA/DI, *A Guide to Monetary Measures,* p. 10. Although growth rates were relatively insensitive to the substitution of one measure for the other, the absolute magnitude of spending on weapons could differ substantially.
46. Testimony before the Becker panel, Nov. 20, 1982, p. 3.
47. Igor Birman, "Professor Holzman, the CIA, Soviet Military Expenditures, and American Security," *Russia,* no. 10 (1984): 43. He cautioned, "in no way must we think that everything is clear to the CIA about the number of Soviet personnel, cannon, and the like" (p. 41). See also Birman, "Size of Soviet Military Expenditures," p. 30.
48. Steven Rosefielde, "The Validity of the CIA's Ruble and Dollar Estimates of Soviet Defense Spending," p. 3. Rosefielde added, however, that intelligence on engineering improvements in weapons "is especially difficult to acquire" and CIA estimates might be "particularly deficient on this score." He gave the same appraisal (without the question regarding the margin of error) to the Becker panel. Testimony before the Becker panel, Feb. 18, 1983, p. 33.

49. William T. Lee, Memorandum for the Record, "Does Alleged Inflation Justify the CIA's Estimates of Soviet Defense Expenditures?" p. 7.
50. Testimony before the Becker panel, Dec. 20, 1982, pp. 5, 8.
51. Lee, *CIA Estimates of Soviet Military Expenditures,* pp. 16–17 and p. 18ff.
52. Ibid., p. 78.
53. *Report of the Working Group,* p. 2. The report did note that CIA and DIA often disagreed on the physical estimates.
54. Berkowitz, et al., "Evaluation of the CIA's Analysis of Soviet Economic Performance," p. 42.
55. The exchanges of information (and the arms control agreements) covered nuclear weapons carriers, not nuclear warheads. Lee claims that the USSR had 45,000 nuclear warheads in the 1980s whereas the intelligence community's estimate was 30,000. Lee, *CIA Estimates of Soviet Military Expenditures,* pp. 16–18.
56. See Rosefielde, "Why the CIA's Estimate of Soviet Defense Procurement Was Off by 200 Percent: The Economic Consequences of Quality Change," pp. 99ff; Rosefielde, "On the Interpretation of Soviet Arms Procurement Expenditures," pp. 46–47. Lee agrees that the dollar costs of Soviet procurement are greatly understated (for strategic missiles by factors of "six or more"). *CIA Estimates of Soviet Military Expenditures,* p. 14.
57. Rosefielde explores this question at length in *False Science,* pp. 59–75. Lee's discussion is shorter, but his conclusion is just as striking: "SCAM's artifacts are largely exempt from the real-cost growth of military technological innovation experience in the West." *CIA Estimates of Soviet Military Expenditures,* p. 15.
58. *Report of the Working Group.* The methodology panel considered the issue in greater detail in its report. It concluded that the cost-estimating equations used by CIA did "take explicit account of intervintage changes in design complexity and the resulting effects on costs." It said, "Professor Rosefielde's published criticism is based on an inadequate understanding of the methodology followed." *Report of the Methodology Panel,* pp. 21–22.
59. Ibid., p. 22.
60. Donald Burton, "A Comparison of the Costs of U.S. and Soviet Tactical Aircraft," pp. 9, 16. In the calculations, the cost of each model was normalized at the same unit of production and a log linear trend line was fitted to the observed costs of each model arranged by initial date of production.
61. Marshall testimony before the Becker panel, Dec. 10, 1982, pp. 3–4.
62. David Ignatius, "Without SALT, the Race Is On," *Washington Post,* June 6, 1986.
63. During David Chu's testimony before the Selin panel, Jan. 13, 1983, p. 5.
64. Jacobsen, "Soviet Military Expenditure," p. 264.
65. Testimony before the Becker panel, Dec. 10, 1982, pp. 4, 7. Thus in the United States even before the end of Selective Service, the amenities

available to U.S. military personnel—ranging from barracks space to off-base housing—were much superior to those in the Soviet Union. Marshall argued that these were necessary to retain U.S. military manpower and should be considered in estimating the dollar cost of maintaining the Soviet forces.

66. Lee was the major exception. On Dec. 20, 1982, he told the Becker panel that O&M defied estimation, but that the CIA estimates were clearly wrong, primarily because of the difficulty of estimating the volume of the activities included in O&M (p. 4). More recently he has said that CIA's O&M building blocks are "samples and estimates of widely varying reliability." He concludes, "Primarily because of incomplete data for the Qs, SCAM's O&M estimates were probably understated, and by large margins." *CIA Estimates of Soviet Military Expenditures,* pp. 23–24.

67. Testimony of Derk Swain and Alan Smith before the Becker panel, Dec. 21, 1982.

68. Lee has vigorously condemned both the old and the new CIA ruble estimates of military RDT&E even though his own estimates are quite similar if his current price estimates are deflated to a 1982 price base. Testimony before the Becker panel, Aug. 29, 1980, pp. 7–8, 16; *CIA Estimates of Soviet Military Expenditures,* p. 24. Lee simply rejects the idea that inflation existed in the RDT&E sector.

69. See David Holloway, *The Soviet Union and the Arms Race,* p. 134.

70. *Report of the Methodology Panel,* p. 26.

71. William T. Lee, *The Estimation of Soviet Defense Expenditures, 1955–1975,* p. 8; Lee, *CIA Estimates of Soviet Military Expenditures,* p. 47; Rosefielde, *False Science,* p. 59ff.

72. The emerging consensus on the power of the defense-industrial complex over development and procurement decisions suggests why caution seemed to be the watchword. Like other industrial enterprises, defense-industrial enterprises were interested in having an achievable state plan that allowed them to meet output and cost targets.

73. For example, Lee advised readers to remember that CIA had not demonstrated that it could estimate U.S. defense expenditures accurately using its methodology. See Lee, "Supplementary Statement to the Subcommittee on General Procurement of the Senate Armed Services Committee," p. 4; and other of his publications.

74. See CIA/DI, DIA, *Dollar Costing.* The discussion of table 6.3 and fig. 6.3 are taken from pp. 31–33 of that paper.

75. Rosefielde thought it "preposterous" for the agency to claim that the procurement estimate was the most stable component of the total estimate. He said that because contractors can't estimate costs of new systems they have not yet produced within 100 percent, "the foolishness of pretending that Soviet procurement can be estimated within a margin of error substantially less than 15 percent should be manifest." Rosefielde's argument is beside the point. He is talking of the failure of American

contractors to project accurately the costs of weapons not yet in production. The estimated dollar costs of Soviet weapons, on the other hand, are based on historical costs of producing weapons of a technological level that exceeded that of almost all Soviet weapons. Rosefielde, "Why the CIA's Estimate of Soviet Defense Procurement Was Off," p. 17.

76. The state budget was an aggregation of the all-union budget, the various union-republican budgets, and the local budgets.
77. Gorbachev gave the new figure in a speech to the Congress of People's Deputies on May 30, 1989. Gorbachev, *Perestroika and Soviet-American Relations*, p. 224. Subsequently, further detail on the revised defense budget was provided by Prime Minister Nikolay Ryzhkov, also in an address to the deputies. *Izvestiya*, June 8, 1989, p. 2.
78. The Institute for Strategic Studies, *The Military Balance;* London, England: IISS (annual) and SIPRI, *World Armaments and Disarmament;* Stockholm, Sweden: SIPRI (annual).
79. Lee makes a small downward adjustment in 1961–64 to take account of his perception that the Soviets moved some procurement into the official defense budget in those years. Lee, *The Estimation of Soviet Defense Expenditures, 1955–1975*, p. 279.
80. William T. Lee, "Trends in Soviet Military Outlays and Economic Priorities, 1970–1988," p. 226.
81. A brief description of the DIA procedure can be found in Maj. Gen. Schuyler Bissel's statement to the JEC. Bissel, "Economic Assessment of the Soviet Union and China," pp. 93–94. Ironically this is the kind of mirror-imaging approach that Graham incorrectly condemned CIA for employing. See note 11 to chapter 2.
82. Some of these efforts include CIA/ORR, *Military Expenditures in the Soviet Budget* (1959 and subsequent years); S. A. Anderson, et al., *Probable Trend and Magnitude of Soviet Expenditures for National Security Purposes;* and Abraham S. Becker, *Soviet National Income, 1958–1964*, pp. 149–56. In "Trends in Soviet Military Outlays," Lee presents his derivation of defense-related budget residuals in some detail.
83. These included net change in public sector inventories, increase in unfinished construction and stocks of equipment requiring installation, and—a particularly difficult estimate—the increase in unfinished construction on collective farms and private agricultural inventories and net additions to nonmilitary state reserves.
84. CIA/DI, *State Reserves and Military Expenditures in the USSR*, and CIA/DI, *Treatment of Defense Outlays in Soviet National Income Statistics*. The agency's research also produced a number of memoranda and working notes on the topic.
85. Peter Wiles, "How Soviet Defense Expenditures Fit into National Income Accounts," in Carl G. Jacobsen, ed., *The Soviet Defense Enigma*. Wiles postulated that procurement entered into the national income accounts net of depreciation, just as investment in the accumulation fund was net of

depreciation and retirements of fixed capital. Gerard Duchene, on the other hand, maintained that it was more likely that procurement was subsumed under the "other expenditures" component of accumulation as a gross figure and that the other components of defense were covered under personal and public consumption. Duchene, "How Much Do the Soviets Spend on Defense," in Carl G. Jacobsen, ed., *The Soviet Defense Enigma*.

86. Given the Soviet propensity to report very few absolute ruble figures (as opposed to index numbers), a great deal of spade work had to be done. See, for instance, J. F. Freeman, "Establishment of an Absolute Ruble Value for the Soviet Concept of 'Gross Industrial Output' "; J. F. Freeman, "A New Source for Data on Soviet Military Expenditure," *Studies in Intelligence* 4, 2 (1962); and William T. Lee, "Value of Soviet Machinery and Military End Item Output: 1950–1958 with Projections to 1965."

87. The derivation of Lee's estimate is set out in his *Estimation of Soviet Defense Expenditures* and his "Trends in Soviet Military Outlays and Economic Priorities, 1970–1988."

88. On Feb. 18, 1983, Rosefielde told the Becker panel that although he found Lee's documentation difficult to assimilate, he supported his general approach, if not his "specific calculations." Later he wrote that weapons estimates derived from "noncivilian machine-building statistics" are more reliable than CIA's "direct-cost series." Parker and Rosefielde, "Soviet Arms Strategy," p. 50. In his own estimate, however, Rosefielde assumes that CIA's estimate of procurement in 1960 (c. 1975) was correct, as was its estimate for 1970 (c. early 1980s). He maintains, however, that there was no significant increase in the prices of military machinery between 1960 and 1970, so all of the increase in the value of military procurement is also an increase in the volume of procurement. He then calculates the average annual percentage increase in procurement in 1961–70 and uses the resulting growth rate to extrapolate the volume of procurement from 1970 to 1979. As a check on his method, Rosefielde compares it with Lee's machine-building residual approach and finds that the alternative methods give consistent results. Rosefielde, *False Science*, pp. 157–96.

89. Bissell, "Economic Assessment," pp. 77–82. This approach attempts to isolate the output of the defense-industrial ministries from total machine-building output. Its weaknesses include the use of official gross output indexes (unadjusted for inflation) and the failure to account for the increasing proportion of production of civil goods in these ministries—up to 40 percent by the end of the 1980s. See also Norbert Michaud, "The Paradox of Current Soviet Military Spending," in Rowen and Wolf, Jr., *The Impoverished Superpower*.

90. The evolution in CIA's views on the machinery purchases residual as a measure of defense procurement can be traced in CIA/ORR Project 14,

4580, *Alternative Measures of Production and Procurement of Military and Space Hardware in the USSR* (unpublished); OER, "The Machinery Purchases Residual as an Indicator of Outlays on Military Hardware by the USSR—New Findings"; OER memorandum, "Comparison of OSR Estimates of Soviet Military Spending with Evidence Provided by Soviet Statistical Reporting"; *Estimating Military Hardware Production from Soviet Industrial Data*, ER RP 74-11, June 1974; CIA/OSR, *Estimating Soviet Spending for Military Hardware From Machine-Building and Metalworking Statistics*, SR M 79-10078, June 1979; and CIA/DI, *Estimating Soviet Military Hardware Purchases: "The Residual" Approach*, SOV 86-10031, June 1986. A version of the last paper by Bonnie K. Matosich appeared in *Gorbachev's Economic Plans*.

91. CIA/OSR, *Estimating Soviet Military Spending*, p. 6.
92. Covering 1966–75, one series was given in 1967 prices using establishment-based statistics; the other two were reported in current prices using establishment statistics in one case and commodity-based statistics in the other.
93. CIA/DI, *Estimating Soviet Military Hardware Purchases*.
94. The most exhaustive treatment of the need for these corrections (and others) can be found in the 1986 paper sourced in note 90 in this chapter, especially pp. 9–13 and appendix.
95. Michael Boretsky, "Comparative Progress in Technology, Productivity, and Economic Efficiency," pp. 231–33.
96. Stanley H. Cohn, *Estimation of Military Durables Procurement Expenditures from Machinery Production and Sales Data*, p. 17. In his report Cohn severely criticized Lee's and Boretsky's competing residual estimates: "both analysts have been tendentious in their presentations and have neglected to introduce the usual cautions, checks, and qualifications demanded by high standards of economic analysis" (p. 2).
97. Daniel L. Bond and Herbert S. Levine, "The Soviet Machinery Balance and Military Durables in SOVMOD," pp. 300–304.
98. Ibid., p. 313.
99. Testimony before the Becker panel, Feb. 3, 1983, pp. 3, 11.
100. *Report of the Working Group*, p. 13.
101. A group of researchers in the CIS Statistical Committee also rejected the official measures of the growth of industrial production. They developed a new sample that showed a fall in industrial output of 9.5 percent in 1991, compared with a 2.5 percent decline reported on the basis of enterprise accounting in "comparable prices." They found the most significant discrepancies in branches with a broad assortment of products that were "renewed" frequently and where the price factor was not always taken into account—especially machine building and light industry. "Opredeleniye pokazateley dinamike promyshlennoy produktsii v uslovikh razvitiya rynochnikh otnoshenii i liberatzii tsen (Determination of Indexes of the

Dynamics of Industrial Production in Conditions of the Development of Market Relations and Price Liberalization)," *Informatzionniy byuleten' statkomiteta SNG,* no. 3 (1992): 11.

102. Ibid. For one thing the CIA index of machinery output incorporates the CIA estimate of defense procurement while Khanin and Eidelman are unclear as to how they deal with its military component of machinery production.

103. We have found only one other reference to the share of defense industry in total machine-building output. A. Ozhegov puts it at 30 percent in Soviet domestic prices. There are, however, numerous statements concerning the share of civilian production in the output of defense industry. Ozhegov, "The Defense Complex: New Economic Problems," *Studies on Soviet Economic Development* 2, 1 (1991): 67. All put this share at 35–40 percent in the late 1980s. The sources include some of the leading lights in the Soviet government and defense industry: Prime Minister Nikolay Ryzhkov (Moscow television service in Russian, 1745 GMT, Aug. 10, 1990, and *Izvestiya,* Dec. 14, 1989); L. A. Voronin, first deputy chairman of the USSR Council of Ministers (*Pravda,* Sept. 26, 1989, p. 3); V. Komorov, a deputy chief of the State Military Industrial Commission of the USSR Council of Ministers (*Pravitelstevnny vestnik,* no. 18, Sept. 1989, p. 11); V. Shabanov, deputy minister of Defense for Armaments (*Krasnaye zvezda,* Aug. 18, 1989, p. 1); and O. D. Bakhlanov, a secretary of the Central Committee of the CPSU (*Pravda,* Dec. 9, 1989, p. 3).

104. Lee can only make his procurement estimate fit by misinterpreting some of the references in the preceding note—i.e., that the shares of civil production in defense industry (unambiguously reported in the sources) actually refer to the share of civilian output in total machinery production. See, for example, "Trends in Soviet Military Outlays," p. B-9; and *Russian Military Expenditures,* pp. 11, 17, 31.

105. Moscow television service in Russian, 1700 GMT, Oct. 9, 1989.

106. Soviet press articles indicated that the 132 billion rubles included about 33 billion rubles in spending under civilian accounts: accounts for foreign economic activity, support of civilian sectors, and revenues and spending attributed to a newly established stabilization fund formed from depreciation funds and excess profits.

107. In this connection, Lt. Gen. Leonid Shebarshin, a retired intelligence official, claimed, "In my time, the whole KGB spent 5 billion a year." *Komsomol'skaya pravda,* Sept. 9, 1993, p. 2.

108. Pressed for more details about the recently released new defense budget, Akhromeyev told an interviewer, "You are young and you want it all at once on a plate. Just think how we established this budget of 77.3 billion rubles. After all we did not have it as such. Previously the armed forces were only interested in what they were given for upkeep, in accordance with a government decision. As far as research and development was concerned, that came under the ministries of defense industry sectors.

Nor did we pay for series production deliveries. First, we had to bring this together." Moscow television service, Oct. 9, 1981.
109. P. Gushvin, "Doveriya k statistike (Trust in Statistics)," *Vestnik statistiki* 9 (Sept. 1992): 7–8.
110. Valentin Pavlov, during his brief tenure as Soviet prime minister, gave a meeting of Moscow workers a figure he said was never published anywhere before: "Usually in the plans, it was always stated that our accumulation fund was 25 or 24 percent [of national income utilized], and all the rest was consumption. It was even stated that personal consumption and public consumption comprised 90 percent of total national income. Well, I in general can say that the so-called accumulation here, including military spending, was always around 36–38 percent. If you evaluate this figure realistically, then you will understand what this means. This is not because somehow they were very malicious there, or that they fail to understand something. But this was the international situation." Moscow Central Television First Program in Russian, 1600 GMT, Mar. 22, 1991.
111. Finance Minister Boris Gostev said the new budget even covered pensions, hospitals, sanatoria, and pioneer camps and that western speculation that it is understated is "groundless" and "misinformation." Moscow Tass in English, 1826 GMT, Jan. 1, 1989. Moiseyev, head of the General Staff, complained that western estimates of Soviet defense spending "continue to dominate information sources in NATO countries." *Pravda*, June 11, 1989, p. 5.
112. Moscow Domestic Service in Russian, 1545 GMT, Apr. 27, 1990; and *Izvestiya*, Apr. 29, 1990, p. 2.
113. Moscow television service in Russian, 1742 GMT, May 7, 1990.
114. Eduard Shevardnadze, *The Future Belongs to Freedom*, p. 54.
115. *Washington Times*, Aug. 1, 1988. Other reporting that puts defense spending in this general area includes economist B. Rayzberg (far more than 100 billion rubles, per *Nedelya*, Feb. 22, 1989); journalist V. Selyunin (about 126 billion rubles, per *Sotsialisticheskaya industriya*, Apr. 6, 1989); parliamentarian V. Lopatin ("a Ministry of Defense representative told us defense spending is about 120 billion rubles a year," per *Nedelya*, May 28, 1990); academician G. Arbatov (about 122 billion rubles, per *Izvestiya*, July 21, 1990); V. M. Zakharov, Feb. 1994, the USSR spent between 15 and 18 percent of GNP for "defense purposes," or 140–170 billion rubles; G. A. Bakharov, Nov. 1993, in 1991 defense spending amounted to 8 percent of GNP according to official data "while in fact it was actually 18 percent"; former Russian Prime Minister Yegor Gaydar, Jan. 1996, 15 percent of GDP.
116. *Novaya vremya*, no. 10, Mar. 2, 1990.
117. *Izvestiya*, June 7, 1990.
118. Tokyo, *Simbor*, June 23, 1990. Academician Oleg Bogomolov at a conference in the United States in spring 1990 signed on to an estimate of Soviet defense spending of more than 200 billion rubles and later

defended his position by referring to his research and research done by a group of people's deputies (probably the Ryzhov statement). Communication from Vladimir Treml, Mar. 24, 1991. Some observers seem to have lost all touch with reality. Thus Valery Dementev maintained that even a "superficial calculation" shows that defense is about 25 percent of Russian GNP and could be more than 50 percent if relevant U.S. budget categories are included. Dementev, "The Future Interrelations between Civilian and Military RDT&E," in Henry S. Rowen, Charles Wolf, Jr., and Jeanne Zlotnick, eds., *Defense Conversion, Economic Reforms, and the Outlook for the Russian and Ukrainian Economies,* p. 193.

119. V. N. Lobov, "Voennaya reforma: istoricheskiye predposylki i osnovniye napreveleniya (Military Reform: Historical Preconditions and Basic Directions)," *Voennoistoricheskiy zhurnal,* no. 11 (1991): 3.

120. Two Soviet generals, however, after reviewing all of the difficulties in putting together the real defense budget, declared, "Only four men practically knew the true picture of the military budget: the General Secretary, the Council of Ministers Chairman, the Minister of Defense, and its Chief of the General Staff." Kulikov and Gareyev, "Some Issues of Military Technical Policy."

121. The Russian defense budget attracts the same sort of suspicion. In mid-1994, Anatoliy Zhuravlev reported that the government in its report on the implementation of the 1993 budget gave a figure of 7,209 billion rubles but that he had found 3,500 billion rubles of "additional expenditure of a military nature" in other budget line items. In other words the announced budget was one-third less than the "real" budget. *Rossiyskaya gazeta,* June 7, 1994, p. 3.

122. *Washington Times,* Aug. 1, 1988.

123. M. L. Shukhgal'ter, *Machino-stroitel'ny kompleks Rossii: trudny perekhod k rynko,* p. 8 and table 2.1.

124. Dmitri Steinberg had another explanation for the relatively low reported share of defense production in machine-building gross output and final output. He concluded that all labor costs (40 percent of the total) in defense industry are excluded from published production and national income statistics. Therefore, he claimed, all previous estimates of Soviet defense production and national income or individual statistics could only determine the material outlays of defense-industrial enterprises. "Estimating Total Soviet Military Expenditures: An Alternative Approach Based on Reconstructed Soviet National Accounts," in Carl J. Jacobsen, ed., *The Soviet Defense Enigma,* pp. 27–28.

125. Igor Birman, *Secret Income of the Soviet State Budget,* and Birman, "Professor Holzman, the CIA, Soviet Military Expenditures, and American Security."

126. Steinberg, *Soviet National Accounting Methods and Practices,* p. 22. R. W. Davies, with access to Soviet archives of the 1930s, concluded that in 1932

and 1933 a "substantial secret extra budget credit" must have been extended to cover much of the actual defense outlays. Davies, "Soviet Military Expenditure and the Armaments Industry, 1929–1933: A Reconsideration," *Europe-Asia Studies* 45, 4 (1993): 580.

127. There are dissenting views. Lee, for example, derided the new "myth" that the Soviets did not know what they spent on defense. He said that the new, revised budget was a bone thrown to the new Supreme Soviet by leaders reluctant to tell their people how large the defense burden really was. While it may have been hard to have a good knowledge of military expenditures, he concluded, "If the Soviets couldn't count, if the command economy never worked, the USSR would not have become a superpower." Lee, *Russian Military Expenditures*, pp. 7–8. Soviet economist Stanislav Menshikov doubted too that the Ministry of Defense or those in power didn't know what they spent on defense. He argued that even if the ministry did not pay for the weapons, it ordered them, participated in planning for their production, and was represented in the Military-Industrial Commission where such questions were decided. Menshikov, *Sovetskaya ekonomika: katostrofa ili katarsis*, p. 278.

## Chapter 7. The Defense Spending Estimates in Perspective

1. In a memorandum prepared for HPSCI but never sent, SOVA complained that the Millar panel "had no experts in defense-economic issues" and that "the Committee chose to limit their interaction with our defense economists to a brief introductory meeting . . . and a single two hour substantive session . . . held after their report was drafted." James E. Steiner, "Defense Economic Aspects of the Millar Report."
2. The chairman of the working group's methodology subpanel did contest the Millar panel's findings. In a letter to HPSCI Chair David McCurdy, Abraham Becker, then the director of the RAND/UCLA Center for Soviet Studies, wrote that the Millar report "offers some questionable judgments related to the estimation of military outlays," especially their "denigration" of the usefulness of calculating the share of defense in GNP and their criticisms of CIA's estimate of this share. Abraham Becker to Hon. Dave McCurdy, Jan. 14, 1992.
3. Interview with Robert M. Huffstuttler, June 1994.
4. See NIE 11-3-68. The NIE 11-3 series from the mid-1960s through the early 1970s offers some good examples of the use of spending analyses in the estimating process.
5. Ironically success in identifying Soviet development programs and unrealistic judgments regarding their prospects contributed to the delay in recognizing the post-1975 slowdown in the growth of the USSR's defense spending.
6. Andrew W. Marshall, director of Net Assessments, Office of the Secretary

of Defense, to Donald Burton, July 1976. At the time Burton was chief of the analytic division in CIA responsible for military-economic analysis of the USSR.
7. Kaplan, *Wizards*, pp. 315–27.
8. Address to the Nation on Strategic Arms Reduction and Nuclear Deterrence, Nov. 22, 1982, *Public Papers of the Presidents, Ronald Reagan, 1982*, pp. 1506–1507.
9. *Public Papers of the Presidents, Jimmy Carter, 1979*, pp. 176, 695.
10. Memorandum for Sayre Stevens (then director of CIA's National Foreign Assessment Center) from A. W. Marshall, director of Net Assessments, Office of the Secretary of Defense, Sept. 27, 1976.
11. And certainly the costs of overhead reconnaissance that gathered intelligence on order-of-battle and production cannot be included as Lee does in reaching his absurdly high estimate of SCAM's cost: "$5 billion to $10 billion underestimating Soviet military outlays." *Washington Post*, May 13, 1995, p. A14.
12. Responding to the Millar report's contention that CIA presented its results in a way that contributed to misinterpretation and misrepresentation, James Steiner said that CIA had "aggressively and continuously warned in both classified and open fora of the potential for misuse of defense spending data." He cited examples from every CIA testimony before the JEC from 1974 to 1986. Steiner, "Defense Economic Aspects of the Millar Report," pp. 7–10.
13. In this connection the intelligence community gave up on the NIE 11-4 series, *Soviet Military Capabilities and Policies, Main Trends in Soviet Military Policy*, because attempts to integrate political, economic, and military analysis in a coherent, agreed view of Soviet policy produced not much more than generalities of the lowest common denominator.
14. Bruce C. Clarke, letter to authors, July 16, 1997.
15. Noel E. Firth, Letter to the Editor, *Washington Post*, Dec. 21, 1994.

## Appendix A. Costing Improvements, 1975–90

1. CIA/DI, *Trends in Soviet Military Manpower*.
2. CIA/NFAC, *Soviet Military Construction Troops*; CIA/OSR, Memorandum for the Record, "New Estimate of Soviet Railroad Troops Manpower"; and CIA/OSR, Memorandum for the Record, "New Manpower Estimate of the Internal Security Troops of the Soviet Ministry of Internal Affairs."
3. CIA/DI, "Outlook for Soviet Military Manpower in a Decade of Shortage," *USSR Monthly Review*, Apr. 1982, pp. 25–29.
4. CIA/DI, *Soviet Military Manpower: Sizing the Force*.
5. CIA/NFAC, *Military Compensation in the Soviet Union*, p. 1. A 1987 OSR memorandum noted pay increases covering various personnel categories in 1972, 1974, 1977, 1980, 1981, and 1983.

6. CIA/OSR, Memorandums for the Record, "New Estimate of Military Clothing Costs" and "New Estimate of Soviet Military Food Costs."
7. CIA/DI, *Soviet Military Finances in a Guards Tank Division.*
8. The 1983 recognition of a flattening out of Soviet defense procurement beginning in the mid-1970s, discussed in chap. 4, is a major exception to this generalization.
9. CIA/DCI, *Soviet Military Production, 1974–1985,* p. 1.
10. CIA/NFAC, *Soviet Military Aircraft Maintenance,* and CIA/NFAC, *Soviet Naval Ship Maintenance.*
11. Not all of the increased costs were allocated to O&M in the ruble estimate of Soviet defense spending. The portion devoted to capital repair and modernization was added to aircraft procurement.
12. CIA/NFAC, *Soviet Military Aircraft Maintenance,* p. 8.
13. In the CIA ruble estimate the capital repair and modernization part of this cost was counted in procurement of naval ships and boats.
14. CIA/SOVA, Memorandum for the Record, "Revised Methodology for Estimating Land Arms O&M Costs."
15. CIA/SOVA, "Changing Soviet Military Operating Practices: The Drive for Resource Savings," p. 5.
16. For the dollar comparisons, a charge for land use was introduced to achieve better comparability with the cost of U.S. programs.
17. For example, a study of the construction costs associated with Soviet ground force facilities examined in detail sixty-eight installations observed in overhead photography. It divided each facility into functional areas, measured the components, and estimated the cost with the help of Soviet construction cost handbooks. CIA/NFAC, *Characteristics and Costs of the Construction of Soviet Ground Force Facilities.*
18. The reassessment is described in CIA/NFAC, *Trends in Soviet Military Construction.*
19. CIA, *Ruble-Dollar Ratios for Construction.*
20. Derk Swain testimony before the Becker panel, Dec. 21, 1982, p. 18.
21. Ibid., p. 20. A large portion of these outlays were later included in papers dealing with an expanded definition of Soviet national security expenditures.
22. William Trimpin, Memorandum to Larry Gershwin, NIO/Strategic Programs, "Additional Costs of Soviet Deep Underground Construction." Prominent Soviet biologist Zhores Medvedev has reported that between 1949 and 1954 construction of a large network of underground scientific and industrial nuclear bases began in the USSR. *Elsevier* in Dutch, Jan. 7, 1995.
23. NIO M76-0121J, pp. 23–25.
24. NIE 11-3/8-76, vol. 1, p. 2.
25. Ibid., p. 12.
26. Lee, *CIA Estimates of Soviet Military Expenditures,* pp. 133–34. To put the 20 billion ruble figure in perspective, it is equal to the average annual

spending on housing construction in the USSR in 1976–80, cited in *Narodnoye khozyaystvo SSSR v 1985 g.*, p. 367. Lee provides no evidence for his estimate but promises to do so in the future (p. 152).

27. CIA/OSR, "New Approaches to Measuring Soviet Military R&D." The methodology then in use relied on published Soviet statistics on spending and employment in science. A year earlier academician Kirillin had told a British audience that total spending on research and development in the USSR was substantially larger than the published figure for science, which had served as a control total for OSR's RDT&E estimate.
28. CIA/OSR, "Trends in R&D Costs for Soviet Strategic Ballistic Missile Systems."
29. CIA/SOVA, "Indicators of Increased Soviet Military Research and Development."
30. CIA/DI, *Soviet Military R&D: Resource Implications of Increased Weapon and Space Systems for the 1980s.*
31. CIA/DI, *Estimating Soviet Military RDT&E Expenditures.*
32. The capture-recapture technique estimates the likely number of facilities missed by two independently derived databases by measuring the overlap between and the unique facilities identified by each base. This approach, originally developed to track the migration of wild animals, has been applied successfully to intelligence problems such as the estimate of the number of Soviet construction battalions. See W. Feller, *An Introduction to Probability Theory and Its Applications*, vol. 1, pp. 41–45.
33. Thus the wage bill was estimated by multiplying the military R&D manpower by the average wage and social insurance deduction for R&D manpower, the materials used in defense R&D by multiplying manpower by an average materials cost per R&D worker (as reported in published data), and so on.

# References

Anderson, S. A., W. T. Lee, I. M. Oakwood, and J. H. Alexander. *Probable Trend and Magnitude of Soviet Expenditures for National Security Purposes.* SRI, SSC-RM 5205-54. February 1969.

Bakharov, G. A. "Defense Production Pricing in the Context of Conversion." *Military Thought,* no. 11 (November 1993): 56–60.

Becker, Abraham S. *The Burden of Soviet Defense: A Political-Economic Essay.* Project Air Force Report. R-2752-AF. October 1981.

———. "The Price Level of Soviet Machinery in the 1960s." *Soviet Studies* 26, 3 (July 1974): 364–79.

———. *Prices of Producers' Durables in the U.S. and USSR in 1955.* Santa Monica, Calif.: RAND Corporation, August 1959.

———. *Soviet Military Outlays since 1958.* RM-3886-PR. Santa Monica, Calif.: RAND Corporation, June 1964.

———. *Soviet National Income, 1958–1964.* R-464-PR. Santa Monica, Calif.: RAND Corporation, August 1969.

Bergson, Abram. *The Real National Income of Soviet Russia since 1928.* Cambridge: Harvard University Press, 1961.

Berkowitz, Daniel M., Joseph S. Berliner, Paul R. Gregory, Susan J. Linz, and James R. Millar. "An Evaluation of the CIA's Analysis of Soviet Economic Performance, 1970–1990." *Comparative Economic Studies* 35, 2 (Summer 1993): 33–48.

Birman, Igor. "The Economic Situation in the USSR," *Russia,* no. 2 (1981): 13–26.

———. "The Way to Slow the Arms Race." *The Washington Post.* October 27, 1980, p. A15.

———. "Professor Holzman, the CIA, Soviet Military Expenditures, and American Security." *Russia,* no. 10 (1984): 35–56.

———. *Secret Income of the Soviet State Budget.* The Hague: Martinus Nijhoff, 1981.

———. "The Size of Soviet Military Expenditures: A Methodological Aspect."

Preliminary version of a paper for American Enterprise Institute Conference, March 1990.

———. "The Soviet Economy: Alternative Views." *Russia,* no. 12 (1986): 60–74.

———. "Statement to the Working Group of the CIA on Soviet Military Expenses." Submitted in Russian on February 3, 1983; reissued in English on April 3, 1983.

Bissell, Maj. Gen. Schuyler. "Economic Assessment of the Soviet Union and China." Statement submitted to the Joint Economic Committee of Congress. *Allocation of Resources in the Soviet Union and China, 1983: Hearings before the Subcommittee on International Trade, Finance, and Security Economics of the Joint Economic Committee of the United States,* pt. 9.

Bogomolov, O. "Comments on the Report 'The Size of Soviet Military Expenditures: A Methodological Text by I. Birman'." Paper presented at American Enterprise Institute Conference, March 1990.

Bond, Daniel L., and Herbert S. Levine. "The Soviet Machinery Balance and Military Durables in SOVMOD." In *Soviet Economy in the 1980s: Problems and Prospects,* pt. 1, *Selected Papers Submitted to the Joint Economic Committee,* pp. 296–318. U.S. Congress, December 31, 1982.

Boretsky, Michael. "Comparative Progress in Technology, Productivity, and Economic Efficiency." In *New Directions in the Soviet Economy: Studies Prepared for the Subcommittee on Foreign Economic Policy,* pt, 2-A, *Economic Performance,* pp. 133–256. U.S. Congress, 1966.

Brown, Harold. "Military Economic Analysis." Memorandum for the Director of Central Intelligence, May 20, 1977.

Burton, Donald F. "Activities of Military-Economic Planning Branch in Fiscal Year 1970." Memorandum for Chief, Programs Analysis Division, June 18, 1970.

———. "A Comparison of the Costs of U.S. and Soviet Tactical Aircraft." Draft working paper, ca. 1981.

———. "Estimating Soviet Defense Spending." *Problems of Communism* (March–April 1983): 85–93, 131–32.

Bush, George. "Statement of Hon. George Bush, Director, Central Intelligence Agency." Testimony before and statement submitted to the Joint Economic Committee of Congress. *Allocation of Resources in the Soviet Union and China, 1976: Hearings before the Subcommittee on Priorities and Economy in Government of the Joint Economic Committee,* pt. 2. U.S. Congress, 1976.

CIA. *The Economic Impact of Soviet Military Spending.* ER IR 75-3. April 1975.

———. *Estimated Defense Spending in Rubles, 1970–1975.* SR 76-1012U. May 1976.

———. Memorandum to Holders of NIE 11-4-65/11-5-65, *Main Trends in Soviet Military Policy/Soviet Economic Problems and Prospects.* August 5, 1965.

———. *The Role of ORR in Economic Intelligence,* pp. 1–3. CIA/RR Project 3-51. August 1, 1951.

———. *Ruble-Dollar Ratios for Construction.* ER 76-10068. February 1976.

———. *Soviet and U.S. Defense Activities, 1970–1979.* Washington, 1980.

———. *Soviet Capital Investment in Nuclear Weapons Storage Facilities 1951–1976.* SR 76-102 64J. October 1976.

———. *Soviet Goals and Expectations in the Global Power Arena.* Memorandum to Holders of NIE 11-4-78, July 7, 1981.

CIA/DCI. *Soviet Military Production, 1974–1985.* NIIM 86-10002. March 1986.

CIA/DDCI. "Comparison Warsaw Pact and NATO Defense Activities, 1976–1986." Memorandum for Deputy Director for Intelligence, November 5, 1987.

CIA/DDI. "Balancing Soviet Resources in 1986–1990." Note to Paul Ericson, C/SOVA/NIG, June 11, 1985.

———. "Costing Soviet Military Programs." Memorandum for the Director of Central Intelligence, April 25, 1972.

———. "Meeting with OSR and OER People, Friday 24 October 1975, 1530 hours, re: Soviet Emigré." Memorandum from Deputy Director for Intelligence, October 24, 1975.

———. Note to Director of Soviet Analysis, May 31, 1983.

———. Note to Director of Soviet Analysis, December 10, 1987.

———. "Secretary of Defense/DCI Discussion of Soviet Defense Expenditures, 20 July 1984." Draft Memorandum for the Record, July 20, 1984.

———. "Soviet Defense Costing Analysis." Memorandum for the Director of Office of Soviet Analysis, DDI #4206-82, May 19, 1982.

———. "Soviet Defense Expenditures." Note to Director, SOVA, April 1, 1985.

———. "Soviet Military-Economic Analysis." Memorandum for Dr. Ivan Selin, Chairman, Military-Economic Advisory Panel, July 29, 1982.

———. "Soviet Procurement/Ruble Costing Paper." Note to Director of Central Intelligence, Deputy Director of Central Intelligence, June 24, 1983.

CIA/DI. *The 1969 Soviet Defense Budget.* SR IM 69-2. January 1969.

———. *The 1970 Soviet Defense Budget in Perspective: Trends in Spending for Defense and Space since 1960.* SR IM 70-1. January 1970.

———. *Changing Soviet Aircraft Maintenance Procedures: Retaining Readiness while Saving Resources.* SOV 90-10043. July 1990.

———. *A Comparison of Soviet and U.S. Defense Activities, 1972–1981.* SOV 83-10035. February 1983.

———. *A Comparison of Soviet and U.S. Defense Activities, 1973–1987.* SOV 88-10052. July 1988.

———. *A Comparison of Soviet and U.S. Defense Activities, 1976–1985.* SOV 86-10028. May 1986.

———. *A Comparison of Warsaw Pact and NATO Defense Activities, 1976–1986.* SOV 87-10077/EUR 87-10032. December 1987.

———. *The Cost of Projected Soviet Strategic Forces.* SOV M 86-20010JX. February 5, 1986.

———. *The Costs of Soviet Involvement in Afghanistan.* SOV 87-10007. February 1987.

———. *Defense's Claim on Soviet Resources.* SOV 87-10011X. February 1987.

———. *Estimated Defense Spending in Rubles, 1970–1975.* SR 76-10121. May 1976.

———. *Estimating Soviet Military Hardware Purchases: The "Residual" Approach.* SOV 86-10031. June 1986.

———. *Estimating Soviet Military RDT&E Expenditures.* SOV 86-10030. July 1986.

———. *Gorbachev's Modernization Program: Implications for Defense.* SOV 86-10015X. March 1986.

———. *A Guide to Monetary Measures of Soviet Defense Activities.* SOV 87-10069. November 1987.

———. *Moscow's Defense Spending Cuts Accelerate.* OSE 92-1006. March 1992.

———. "Outlook for Soviet Military Manpower in a Decade of Shortage." *USSR Monthly Review* (April 1982): 25–29.

———. *The Price of Strength: Broader Soviet Force Goals Driving Up Defense Spending.* SR IM 68-6. February 1968.

———. *Projecting Soviet Military Forces and Weapons Procurement.* SOV 87-10066. November 1987.

———. "Sec. Def./DCI Discussion of Soviet Defense Expenditures, 20 July 1984." Memorandum for the Record, July 25, 1984. Signed by Lt. Gen. James A. Williams, Director, Defense Intelligence Agency, and Robert M. Gates, Deputy Director for Intelligence, Central Intelligence Agency.

———. *Soviet Defense Expenditures, 1963–1972.* SR IM 72-7. March 1972.

———. *Soviet Defense Expenditures and Their Economic Impact through 1970.* CIA/RR MR 64-1. December 1964.

———. *The Soviet Defense Industry: Coping with the Military-Technological Challenge.* SOV 87-10035 DX. July 1987.

———. *Soviet Defense Spending: Recent Trends and Future Prospects.* SOV 83-10135 CX. July 1983.

———. *Soviet Economic Problems and Prospects.* A study prepared for the use of the Subcommittee on Priorities and Economy in Government of the Joint Economic Committee, U.S. Congress, 1977.

———. *Soviet Expenditures for Defense and Space Programs, 1962–1971.* SR IM 71-5. March 1971.

———. *Soviet Gross National Product in Current Prices, 1960–1980.* SOV 83-10037. March 1983.

———. *Soviet Measures to Reduce the Vulnerability of the Economy in a Nuclear War.* SOV 83-10066. April 1983.

———. *Soviet Military Finances in a Guards Tank Division.* SOV M 89-20047X. August 15, 1989.

———. *Soviet Military Manpower: Sizing the Force.* SOV 90-10046X. August 1990.

———. *Soviet Military R&D: Resource Implications of Increased Weapon and Space Systems for the 1980s.* SOV 83-10064. April 1983.

———. *Soviet Spending for Defense: An Annual Review,* vol. 1, *Trends in Ruble Expenditures.* SR IR 73-11. August 1973.

———. *Soviet Spending for Defense: An Annual Review,* vol. 2, *A Monetary Comparison of Soviet and U.S. Defense Activities.* SR IR 73-12. August 1973.

———. *State Reserves and Military Expenditures in the USSR*. ER IR 69-25. September 1969.
———. *Treatment of Defense Outlays in Soviet National Income Statistics*. ER IR 71-021. July 1971.
———. *Trends in Soviet Military Manpower*. SR 77-10103. September 1977.
———. *USSR: Economic Projections, 1982–1990*. SOV 82-10127. September 1982.
———. *USSR: Economic Projections through 1990—A New Look*. SOV 84-10017. March 1984.
CIA/DI, DIA. *Dollar Costing of Foreign Defense Activities: A Primer on Methodology and Use of the Data*. SOV 88-10048. July 1988.
CIA/Military-Economic Advisory Panel. *First Report—Military Economic Review Panel*. July 1, 1974.
———. *Military Economic Advisory Panel Meeting of April 13–16, 1978*. Submitted to DCI Stansfield Turner. June 8, 1978.
———. *Report of the Military-Economic Advisory Panel to Admiral Stansfield Turner, Director of Central Intelligence*. August 1977.
———. *Second Annual Report to the Deputy Director for Intelligence*. January 22, 1976.
CIA/NFAC. *Characteristics and Costs of the Construction of Soviet Ground Force Facilities*. SR 81-10019/IS 81-10038. March 1981.
———. *The Development of Soviet Military Power: Trends since 1965 and Prospects for the 1980s*. SR 81-10035X. April 1981.
———. *Estimated Soviet Defense Spending: Trends and Prospects*. SR 78-10121. June 1978.
———. *Military Compensation in the Soviet Union*. SR 81-10003. January 1981.
———. *Simulations of Soviet Growth Options to 1985*. ER 79-10131. March 1979.
———. *Soviet and U.S. Defense Activities, 1971–1980: A Dollar Cost Comparison*. SR 81-10005. January 1981.
———. *Soviet Military Aircraft Maintenance*. SR 79-10141. October 1979.
———. *Soviet Military Construction Troops*. SR 80-10095. July 1980.
———. *Soviet Naval Ship Maintenance*. SR 81-10048. April 1981.
———. *Trends in Soviet Military Construction*. SR 81-10001. January 1981.
CIA/NIC. "The Controversy Over 'Soviet Defense Spending'." Memorandum for Chairman, National Intelligence Council from National Intelligence Officer at Large, March 9, 1983.
CIA/OER. "Comparison of OSR Estimates of Soviet Military Spending with Evidence Provided by Soviet Statistical Reporting." August 26, 1968.
———. *Estimating Soviet Military Hardware Production from Soviet Industrial Data*. ER RP 74-11. June 1974.
———. "The Machinery Purchases Residual as an Indicator of Outlays on Military Hardware by the USSR—New Findings." August 23, 1967.
———. *Response to Questions from the Proxmire Briefing*. ER M 76-10385. June 9, 1976.
CIA/ORR. *1955 Ruble-Dollar Ratio for Cost of Construction of the Soviet "SKORYY" Class Destroyer*, CIA/RR RA-24. January 31, 1958.

References

———. "Alternative Measures of Production and Procurement of Military and Space Hardware in the USSR." ORR Project 14.4580. August 27, 1965.

———. "Contribution to NIE 11-5-55, Air Defense Capabilities of the Sino-Soviet Bloc, 1955–1960." CIA/RR IP-394 (revised). June 10, 1955.

———. *Cost of Construction of the Soviet W-Class Submarine*, CIA/RR RA 59-19. December 1959.

———. *Estimates of Costs in Rubles of Building Various Types of Merchant Ships in the USSR.* CIA/RR IM-448. February 28, 1957.

———. *Military Expenditures in the Soviet Budget, Selected Years, 1950–1957.* CIA/RR RA 59-6. April 1959.

———. *Military-Economic Programs of the USSR, 1947–1961.* SC no. 01819/57. March 14, 1957.

———. "ORR Contribution to NIE 11-4-66, *Main Trends in Soviet Military Policy.*" CIA/RR MP 66-1. April 1966.

———. *Pay for Personnel of the Soviet Ground Forces, 1957.* CIA/RR SC 59-15. October 16, 1959.

———. *Ruble-Dollar Ratio for Soviet Aircraft.* CIA/RR A. ERA 60-9. August 1960.

———. "Soviet Defense Expenditures, 1955–1964." Contribution to the Memorandum to Holders of NIE 11-4-65, *Main Trends in Soviet Military Policy.* CIA/RR MP 65-1. June 2, 1965.

———. *Soviet Military Expenditures, 1958–1965.* CIA/RR ER SC 61-4. April 28, 1961.

———. *Soviet Military Expenditures by Major Missions.* CIA/RR ER 61-15. April 1961.

———. *Theory and Application of the Learning Curve.* CIA/RR RA-7. July 24, 1956.

CIA/OSR. *The Economics of Soviet Military Aircraft Maintenance.* SR M 79-10032JX. March 30, 1979.

———. *Estimating Soviet Spending for Military Hardware from Machine-Building and Metalworking Statistics.* Working paper. SR M 79-10078. June 1979.

———. *Methods for Estimating Soviet Defense Expenditures.* SR M 77-10020C. February 1977.

———. "Monthly Activity Report for March 1976—Programs Analysis Division." April 26, 1976.

———. "NATO-Warsaw Pact Defense Cost Comparisons." Memorandum for Director, Strategic Research, April 3, 1979.

———. "New Approaches to Measuring Soviet Military R&D." Memorandum for Deputy Director for Intelligence, January 18, 1976.

———. "New Estimate of Soviet Military Clothing Costs." Memorandum for the Record, January 21, 1980.

———. "New Estimate of Soviet Military Food Costs." Memorandum for the Record, February 20, 1980.

———. "New Estimate of Soviet Railroad Troops Manpower." Memorandum for the Record, September 11, 1980.

———. "New Manpower Estimate of the Internal Security Troops of the Soviet Ministry of Internal Affairs." Memorandum for the Record, July 18, 1980.

———. "O&M General Purpose Vehicles and Other Hardware Items." Typescript, Appendix 7-A. Spring 1976.

———. "OSR Estimates of Soviet Defense Spending for Military RDT&E and Total RDT&E Spending." Memorandum for the Record, December 21, 1979.

———. "OSR's 1977 Projection." Undated briefing paper.

———. "OSR's Analytical Resources Assignments by Topic, Category, and Country." Memorandum for Director, National Foreign Assessment Center, December 13, 1979.

———. "Price Base for the Military Expenditures Series." Memorandum for Chief Programs Analysis Division, OSR; Deputy Chief Programs Analysis Division, OSR, May 16, 1969.

———. "A Reduction in Soviet Military Production?" Internal Memorandum, May 23, 1977.

———. "SCAM-Update-Utility Costs." Memorandum for the Record, January 8, 1976.

———. "Soviet Military Construction Programs Changing." *Strategic Intelligence Monthly Review* (September 1980): 3–5.

———. "Soviet Pre-induction Military Training Costs, in Dollars; January 1976 SCAM Update." Memorandum for the Record, January 27, 1976.

———. "Transmittal of Information to Congressman Aspin." Memorandum for the Record, April 25, 1975.

———. "Trends in R&D Costs for Soviet Strategic Ballistic Missile Systems." *Strategic Intelligence Monthly Review* (September 1981): 1–4.

CIA/OSR and OER. "Paper on the USSR's Defense Burden." Memorandum for Deputy Director of Intelligence, May 28, 1974.

CIA/OSR/ME. "Revisions to Soviet Military Construction Cost Estimate." Memorandum for the Record, January 16, 1980.

CIA/SOVA. "Changing Soviet Military Operating Practices: The Drive for Resource Savings." *USSR Monthly Review,* (June 1988).

———. "Chronology of Events on the Subject of Soviet/U.S. Defense Comparisons." Memorandum for Chief, Econometric Analysis Division, August 10, 1984.

———. "CIA's Dollar Cost Work: Where Do We Go From Here?" Memorandum for Director of Soviet Analysis, August 3, 1990.

———. "Dollar Cost Analysis of Soviet Defense Activities." Memorandum for Director of Central Intelligence, Deputy Director of Central Intelligence, December 22, 1981.

———. "Estimating Soviet Defense Spending in Current Prices." Memorandum for the Record, May 3, 1989.

———. "Indicators of Increased Soviet Military Research and Development." *USSR Monthly Review* (October 1982): 17–20.

———. "The Leveling Off of Soviet Defense Procurement Spending and Recent SOVA Research." Memorandum for the Record, October 10, 1986.

———. "Meeting with Norb Michaud, Chief, DB-4E, on Terms of an Agreement on Defense Cost Comparisons." Memorandum for Director of Soviet Analysis, September 10, 1984.

———. "Meeting with Richard Kaufman, Associate Staff Director, Joint Economic Committee." Memorandum for the Record, October 19, 1984.

———. "NIE 11-3/8: Force Projections." Memorandum for Deputy Director for Intelligence, from Douglas J. MacEachin, Director of Soviet Analysis, April 22, 1986.

———. "Our Comparison of NATO and Warsaw Pact Defense Activities." Memorandum for Deputy Director for Intelligence, November 20, 1987.

———. "Overview: The Impact of the New Five-Year Plan on the Defense Sector." May 1986.

———. "The Pay Model: Position Pay for Warrant Offices and Reenlisted Personnel, Explained." Memorandum for the Record, September 23, 1987.

———. "Recommended Response to Ikle Request." Memorandum for Deputy Director for Intelligence, January 31, 1985.

———. "Record of Published Estimates of Soviet Defense Spending." Memorandum for Deputy Director for Intelligence, January 25, 1985.

———. "The Release of Dollar Cost Comparisons of U.S. and Soviet Defense Costs." Memorandum for Deputy Director for Intelligence, December 13, 1984.

———. "Request for Warsaw Pact–NATO Defense Comparisons." Memorandum for Deputy Director for Intelligence, January 5, 1988.

———. "Review of Ruble Procurement Prices." Memorandum for the Record, November 8, 1990.

———. "Revised Methodology, for Estimating Land Arms O&M Costs." Memorandum for the Record, October 14, 1987.

———. "Revision of Estimated USSR Premilitary Training Costs." Memorandum for the Record, July 18, 1984.

———. "SOVA's Estimate for Military Machinery Production." Memorandum for Director of Soviet Analysis, May 25, 1982.

———. "Soviet Defense Costs." Memorandum for Deputy Director for Intelligence, June 3, 1983.

———. "The State of the Soviet Economy." Memorandum for the Deputy Director for Intelligence, April 22, 1987.

———. "Talking Points for the DDI on Curtailing Our Work on Estimating Dollar Cost Comparisons of U.S.-Soviet Defense Activities." Memorandum for the Deputy Director for Intelligence, August 16, 1990.

———. "Weinberger Briefing on Soviet Defense Spending Issues." Memorandum for Deputy Director for Intelligence, April 11, 1985.

Coffin, David. *Development of Economic Intelligence*, vol. I, *1950–1960*, p. 3. OER 1. September 1973.

Cohn, Stanley H. *Estimation of Military Durables Procurement Expenditures from Machinery Production and Sales Data.* Informal Note SSC-IN-78-13. Arlington, Va.: SRI International Strategic Studies Center, September 1978.

Committee on the Present Danger. *Russian Military Expenditures,* Washington, D.C., April 24, 1991.

Davies, R. W. "Soviet Military Expenditure and the Armaments Industry, 1929–1933: A Reconsideration." *Europe-Asia Studies* 45, 4 (1993): 577–608.

Dementev, Valery A. "The Future Interrelations between Civilian and Military RDT&E." In Henry S. Rowen, Charles Wolf, Jr., and Jeanne Zlotnick, eds., *Defense Conversion, Economic Reforms, and the Outlook for the Russian and Ukrainian Economies.* New York: St. Martin's Press, 1994.

DIA. "Dollar Estimates." Memorandum for the Deputy Director for Intelligence, CIA, January 14, 1985.

Director of Central Intelligence Directive (DCID) 15/1. *Production and Coordination of Foreign Economic Intelligence.* September 14, 1954.

Duchene, Gerard. "How Much Do the Soviets Spend on Defense?" In Carl G. Jacobsen, ed., *The Soviet Defense Enigma,* pp. 95–114. Oxford: Oxford University Press, 1987.

Epstein, David F. "The Economic Cost of Soviet Security and Empire." In Henry Rowen and Charles Wolf, Jr., eds., *The Impoverished Superpower: Perestroika and the Soviet Military Burden.* San Francisco: Institute for Contemporary Studies, 1990.

*An Evaluation of CIA Work on Soviet Economic Capabilities, Problems, and Prospects.* Prepared by a panel established by the President's Foreign Intelligence Advisory Board. July 7, 1983.

Fal'tsman, Vladimir. "The New Role of Defense R&D in the Russian Economy." In Henry S. Rowen, Charles Wolf, Jr., and Jeanne Zlotnick, eds., *Defense Conversion, Economic Reform and the Outlook for the Russian and Ukrainian Economies.* New York: St. Martin's Press, 1994.

Feller, W. *An Introduction to Probability Theory and Its Applications, 2nd ed.,* vol. 1, pp. 41–45. New York: Wiley, 1957.

Firth, Noel E. "Letter to the Editor," *Washington Post,* December 21, 1994, as cited in "Soviet Defense Spending Controversy," *Studies in Intelligence* 35, 4 (Winter 1991): 83–87.

Fischer, Ben B. *A Cold War Conundrum: The 1983 Soviet War Scare,* CIA Center for the Study of Intelligence, September, 1997.

Freeman, J. F. "Establishment of an Absolute Ruble Value for the Soviet Concept 'Gross Industrial Output'." ORR CSM no. 66/59. December 2, 1959.

———. "A New Source for Figures on Soviet Military Output." *Studies in Intelligence* 6, 2: 19–26.

Freeman, Lawrence. *U.S. Intelligence and the Soviet Strategic Threat.* Princeton, N.J.: Princeton University Press, 1986.

Gams, E., and V. Makarenko. "Razmyshleniya o raskodakh obshestva na oboronu, voennom byudzhatc i byudzhete ministerstve oborony

(Reflections on Society's Outlays on Defense, the Military Budget and the Budget of the Ministry of Defense)." *Voprosy ekonomiki*, no. 10 (October 1990): 149–53.

Garthoff, Raymond L. *Détente and Confrontation*. Washington, D.C.: The Brookings Institution, 1985.

———. *The Great Transition*. Washington, D.C.: The Brookings Institution, 1994.

Gates, Robert M. *From the Shadows*. New York: Simon and Schuster, 1996.

———. "Soviet Defense Costing Effort." Memorandum for the Director and Deputy Director of Central Intelligence, June 26, 1984.

———. "Statement of Robert Gates, Deputy Director for Intelligence, Central Intelligence Agency." *Allocation of Resources in the Soviet Union and China, 1984: Hearings before the Subcommittee on International Trade, Finance, and Security Economics of the Joint Economic Committee*. U.S. Congress, pt. 10.

——— and Lawrence K. Gershwin. "Soviet Strategic Force Developments." Testimony before a Joint Session of the Subcommittee on Strategic and Theater Nuclear Forces of the Senate Armed Services Committee and the Defense Subcommittee of the Senate Committee on Appropriations, June 26, 1985.

Gaydar, Yegor. "Interview with Yegor Gaydar, Chairman of the Democratic Choice of Russia Party." *Segodnya*, January 17, 1996, p. 5.

Gelman, Harry. *The Brezhnev Politburo and the Decline of Détente*. Ithaca and London: Cornell University Press, 1984.

Godaire, J. G. "The Claim of the Soviet Military Establishment." *Dimensions of Soviet Economic Power*, pt. 1, *The Policy Framework*, pp. 35–46. Materials prepared for the Joint Economic Committee. U.S. Congress, 1962.

Gorbachev, Mikhail. *Perestroika and Soviet-American Relations*. Madison, Conn.: Spine Press, Inc., 1990.

Graham, Daniel. "Comments on CIA Paper, 'The Measurement and Meaning of Defense Burden in the Soviet Setting'." Memorandum for Deputy Director, Central Intelligence Agency, January 28, 1975.

Greenslade, Rush V. "The Many Burdens of Defense in the Soviet Union." *USSR Monthly Review* (April 1982): 57–63.

Guchmazov, S. "Uvazhemiy Eduard Sergeyevich (Dear Eduard Sergeyevich)." *Voprosy ekonomiki*, no. 10 (October 1990): 154–55.

Gushvin, P. "Doveriya k statistike (Trust in Statistics)." *Vestnik statistiki* 9 (September 1992): 3–31.

Hanson, Philip. "Estimating Soviet Defense Expenditure." *Soviet Studies* 30, 3 (July 1978): 403–10.

Heleniak, Timothy. "Puzzling Soviet Labor Force Statistics: Declining State Sector Employment and Employment in the Ministry of Defense." Center for International Research, Bureau of the Census, December 1991.

Helgerson, John L. *CIA Briefings of Presidential Canidates, 1952–1992*. Washington, D.C.: CIA Center for the Study of Intelligence, n.d.

Helms, Sen. Jesse. "Why Has U.S. Intelligence Underestimated Soviet Defense Spending for Thirty Years?" *Estimating the Size and Growth of the Soviet*

*Economy: Hearing before the Committee on Foreign Relations of the United States Senate*, pp. 99–113. 101st Cong., 2nd sess., July 16, 1990.

Helms, Richard. "Military-Economic Intelligence." Memorandum for Lt. General Donald V. Bennett, USA, Director, Defense Intelligence Agency, April 25, 1972.

Holloway, David. *The Soviet Union and the Arms Race*. New Haven: Yale University Press, 1983.

Holzman, Franklyn D. "Are the Soviets Really Outspending the U.S. on Defense?" *International Security* 4, 4 (Spring 1980): 86–104.

———. "CIA Estimates of Soviet Military Spending: The Author Replies." *International Security* 14, 4 (Spring 1990): 193–98.

———. "How CIA Concocts Soviet Defense Numbers." *New York Times*, October 25, 1989, p. A30.

———. "Myths that Drive the Arms Race." *Challenge* (September–October 1984): 32–36.

———. "Politics and Guesswork: CIA and DIA Estimates of Soviet Military Spending." *International Security* 14, 2 (Fall 1989): 101–31.

———. "Politics, Military Spending, and the National Welfare." *Comparative Economic Studies* 36, 3 (Fall 1994): 1–4.

———. "Soviet Military Spending: Assessing the Numbers Game." *International Security* 6, 4 (Spring 1982): 78–101.

Ignatius, David. "Without SALT, the Race Is On." *Washington Post*, June 6, 1986, p. F 5.

International Institute for Strategic Studies. *The Military Balance, 1965–66*. London.

International Institute for Strategic Studies. *The Military Balance, 1967–68*. London.

International Institute for Strategic Studies. *The Military Balance, 1969–70*. London.

International Institute for Strategic Studies. *The Military Balance, 1970–71*. London.

International Institute for Strategic Studies. *The Military Balance, 1971–72*. London.

International Institute for Strategic Studies. *The Military Balance, 1972–73*. London.

International Institute for Strategic Studies. *The Military Balance, 1973–74*. London.

International Institute for Strategic Studies. *The Military Balance, 1974–75*. London.

International Institute for Strategic Studies. *The Military Balance, 1975–76*. London.

International Institute for Strategic Studies. *The Military Balance, 1976–77*. London.

International Institute for Strategic Studies. *The Military Balance, 1977–78*. London.

**References**

International Institute for Strategic Studies. *The Military Balance, 1978–79*. London.

International Institute for Strategic Studies. *The Military Balance, 1979–80*. London.

International Institute for Strategic Studies. *The Military Balance, 1980–81*. London.

International Institute for Strategic Studies. *The Military Balance, 1981–82*. London.

International Institute for Strategic Studies. *The Military Balance, 1982–83*. London.

International Institute for Strategic Studies. *The Military Balance, 1983–84*. London.

International Institute for Strategic Studies. *The Military Balance, 1984–85*. London.

International Institute for Strategic Studies. *The Military Balance, 1985–86*. London.

International Institute for Strategic Studies. *The Military Balance, 1986–87*. London.

International Institute for Strategic Studies. *The Military Balance, 1987–88*. London.

International Institute for Strategic Studies. *The Military Balance, 1988–89*. London.

International Institute for Strategic Studies. *The Military Balance, 1989–90*. London.

International Institute for Strategic Studies. *The Military Balance, 1991–92*. London.

Jacobsen, Carl G. "Soviet Military Expenditure and the Soviet Defense Burden." In *World Armaments and Disarmament, SIPRI Yearbook 1986*, pp. 263–74. Stockholm International Peace Research Institute. Oxford University Press, Oxford, England, 1986.

Kaplan, Fred. *The Wizards of Armageddon*. New York: Touchstone, 1984.

Kent, Sherman. Memorandum for Deputy Director for Intelligence, January 10, 1963.

Khanin, G. I. *Dinamika ekonomicheskovo razvitiya SSR (Dynamics of the Economic Development of the USSR)*. Novosibirsk: Nauka, 1991.

Kudrov, V. "Are the Calculations of Economic Growth Rates for the USSR and Russia Reliable?" *Problems of Economic Transition* (October 1994): 53–66.

Kulikov, Viktor G., and Makhmut A. Gareyev. "Some Issues of Military Technical Policy." *Vooruzheniye, politika, konversiya*, no. 2 (1993): 29–33.

Kushnirsky, Fyodor I. "Methodological Aspects in Building Soviet Price Indices." *Soviet Studies* 37, 4 (October 1985): 505–19.

Lee, William T. *CIA Estimates of Soviet Military Expenditures—Errors and Waste*. Washington, D.C.: American Enterprise Institute, 1995.

———. "Does Alleged Inflation Justify the CIA's Estimates of Soviet Defense Expenditures?" Memorandum for the Record, May 8, 1981.

———. *The Estimation of Soviet Defense Expenditures, 1955–1975*. New York: Praeger Publishers, 1977.
———. "The Real Implications of the CIA/DIA Joint Statement to the JEC." Memorandum, July 1, 1986.
———. "Reply to CIA Criticism of My Estimates of Soviet Defense Expenditures—An Open Letter." Typescript, August 29, 1980.
———. "Soviet Defense Expenditures in the 10th FYP." *Osteuropa Wirtschaft* 4 (December 1977): 273–92.
———. "Statement of William T. Lee, Defense Intelligence Agency." *Estimating the Size and Growth of the Soviet Economy: Hearing before the Committee on Foreign Relations of the United States Senate,* 101st Cong., 2nd sess., July 16, 1990.
———. "Supplementary Statement to the Subcommittee on General Procurement of the Senate Armed Services Committee." November 8, 1979.
———. "Team B Soviet SDI Response Cost Effectiveness." Memorandum, March 1990.
———. "Trends in Soviet Military Outlays and Economic Priorities, 1970–1988." Appendix to *Estimating the Size and Growth of the Soviet Economy: Hearing before the Committee on Foreign Relations of the United States,* pp. 98–258. 101st Cong., 2nd sess., July 16, 1990.
———. "Value of Soviet Machinery and Military End Item Output 1950–1958 with Projections to 1965." Draft ORR project. July 1959.
Leggett, Robert E., and Sheldon T. Rabin. "A Note on the Meaning of the Soviet Defense Budget." *Soviet Studies* 30, 4 (October 1978): 357–66.
Ligachev, Yegor. *Inside Gorbachev's Kremlin*. Translated by Catherine A. Fitzpatrick, Michele A. Berdy, and Dobrechna Dyrcz-Freeman. New York: Pantheon Books, 1993.
Lobov, V. N. "Voennaya reforma: istoricheskiye predposylki i osnovniye napreveleniya (Military Reform: Historical Preconditions and Basic Directions)." *Voenno-istoricheskiy zhurnal,* no. 11 (1991): 2–10.
MacEachin, Douglas. "Gorbachev's Economic Strategy and Soviet Defense Spending." Memorandum, December 13, 1985.
Maresca, John J. "Study on NATO-Warsaw Pact Defense Activities, 1976–1985." Memorandum for Director, Office of Soviet Analysis, CIA, and Director, Office of European Analysis, CIA, December 23, 1986.
Marshall, Andrew W. Memorandum for Sayre Stevens, Director of National Foreign Assessments Center, September 27, 1976.
———, J. E. Loftus, and G. E. Pugh. *Project Lamp, Systems Analysis, and the Military Estimates Process.* April 3, 1961.
———. "Commentary." In *Gorbachev's Economic Plans,* vol. 1, *Study Papers,* pp. 481–84. Submitted to the Joint Economic Committee, U.S. Congress, 1987.
Matosich, Bonnie K. "Estimating Soviet Military Hardware Purchases: The Residual Approach." In *Gorbachev's Economic Plans,* vol. 1., *Study Papers,* pp. 431–61. Submitted to the Joint Economic Committee, U.S. Congress, 1987.
McGwire, Michael. *Perestroika and National Security.* Washington, D.C.: The Brookings Institution, 1991.

Menshikov, Stanislav. *Sovetskaya ekonomika: katostrofa ili katarsis (Soviet Economy: Catastrophe or Catharsis)*. Moscow: Inter-verso, 1990.

Meyer, Herbert. "Fred Ikle's Concern about Intelligence Community Soviet Cost Estimates." Memorandum for Director of Central Intelligence, Deputy Director of Central Intelligence, and Chairman, National Intelligence Council, January 19, 1984.

Meyer, Stephen M. "Economic Constraints in Soviet Military Decision-Making." In Henry S. Rowen and Charles Wolf, Jr., eds., *The Impoverished Superpower*, pp. 201–20. San Francisco: ICS Press, 1990.

Michaud, Norbert. "The Emigré Who Raised the Burden." *Studies in Intelligence* 35, 2 (Summer 1991): 57–61.

———. "The Paradox of Current Soviet Military Spending." In Henry S. Rowen and Charles Wolf, Jr., eds., *The Impoverished Superpower*, pp. 111–26. San Francisco: ICS Press, 1990.

National Security Council Intelligence Directive (NSCID). No. 3, January 13, 1948.

NIE-3. *Soviet Capabilities and Intentions*. November 15, 1950.

NIE-64. *Soviet Bloc Capabilities through Mid-1953*, pt. 1. November 12, 1952.

NIE-65. *Soviet Bloc Capabilities through 1957*. June 15, 1953.

NIE 11-4-54. *Soviet Capabilities and Probable Courses of Action through Mid-1959*. September 14, 1954.

NIE 11-6-54. *Soviet Capabilities and Probable Programs in the Guided Missile Field*. October 5, 1954.

NIE 11-5-55. *Bloc Air Defense Capabilities, 1955–1960*. July 12, 1955.

NIE 11-4-56. *Soviet Capabilities and Probable Courses of Action through 1961*. August 2, 1956.

NIE 11-4-60. *Main Trends in Soviet Capabilities and Policies, 1960–1965*. December 1, 1960.

NIE 11-4-63. *Soviet Military Capabilities and Policies, 1962–1967*. March 22, 1963.

NIE 11-4-65. *Main Trends in Soviet Military Policy*. April 14, 1965.

NIE 11-3-68. *Soviet Strategic Air and Missile Defenses*. October 31, 1968.

NIE 11-3/8-76. *Soviet Forces for Intercontinental Conflict through the Mid-1980s*, vol. 1, *Key Judgments and Summary*. December 21, 1976.

NIE 11-3/8-83. *Soviet Capabilities for Strategic Nuclear Conflict, 1983–1993*. March 6, 1984.

NIE 11-3/8-86 II. *Soviet Forces and Capabilities for Strategic Nuclear Conflict through the Mid-1990s*, vol. 2, *The Estimate*. April 24, 1986.

NIE 11-3/8-87 JX/I. *Soviet Forces and Capabilities for Strategic Nuclear Conflict through the Late 1990s*, vol. 1, *Key Judgments and Summary*. July 10, 1987.

NIE 11-3/8-87 JX/II. *Soviet Forces and Capabilities for Strategic Nuclear Conflict through the Late 1990s*, vol. 2, *The Estimate*. July 10, 1987.

NIO M 76-0121J. *Soviet Strategic Objectives: An Alternative View, Report of Team "B."* December 1976.

Noren, James H. "The Russian Military-Industrial Sector and Conversion." *Post-Soviet Geography* 35, 9 (November 1994): 475–521.

Odom, William. "The Riddle of Soviet Military Spending." *Russia* 3 (1981): 53–58.

"Opredeleniye pokazateley dinamike promyshlennoy produktsii v uslovikh razvitiya rynochnikh otnoshenii i liberatzii tsen (Determination of Indexes of the Dynamics of Industrial Production in Conditions of the Development of Market Relations and Price Liberalization)." *Informatzionniy byuleten' statkomiteta SNG*, no. 3 (1992): 7–12.

Organization of the Joint Chiefs of Staff. *United States Military Posture for FY 1983*. Washington, 1982.

Ozhegov, A. "The Defense Complex: New Economic Problems." *Studies on Soviet Economic Development* 2, 1 (1991): 66–73.

Parker, Patrick J., and Steven Rosefielde. "Soviet Arms Procurement Strategy in the Eighties: Conflicting Perceptions of the Soviet Arms Buildup." *Russia* 12 (1986): 49–59.

Pavlov, V. S. "Uvazhemiy Eduard Sergeyivich (Dear Eduard Sergeyivich)." *Voprosy ekonomiki*, no. 10 (October 1990): 156–57.

Pozharov, A. I. "Defense Sufficiency: Military-Economic Aspect." *Military Thought-Voyennaya Mysl'* 7 (1993): 29–35.

Prados, John. *The Soviet Estimate*. New York: The Dial Press, 1982.

Proctor, Edward W. Memorandum for Acting Assistant Director, ORR, June 18, 1964.

———. "Your Comments on 'The Measurement and Meaning of Defense Burden in the Soviet Setting'." Memorandum for Lieutenant General Daniel O. Graham, USA, Director, Defense Intelligence Agency, February 14, 1975.

*Public Papers of the Presidents, Jimmy Carter, 1979*. Washington: U.S. Government Printing Office, 1980.

*Public Papers of the Presidents, Ronald Reagan, 1982*. Washington: U.S. Government Printing Office, 1983.

*Report of the Soviet Working Group*. Prepared at request of Robert Gates, Deputy Director for Intelligence, CIA. May 1985

Rosefielde, Steven. *False Science: Underestimating the Soviet Arms Buildup*. New Brunswick and London: Transaction Books, 1982.

———. "On the Interpretation of Soviet Arms Procurement Expenditures under Conditions of Rapid Technical Progress." *Osteuropa Wirtschaft* 25, 1 (1980):41–53.

———. "Soviet Defense Spending: The Contribution of the New Accountancy." *Soviet Studies* 42, 1 (January 1980): 59–80.

———. "The Strong Separability of Defense and Civilian Growth Potential: A Reassessment of Soviet Growth Prospects in the 1980s." Draft paper. February 1983.

———. "The Validity of the CIA's Ruble and Dollar Estimates of Soviet Defense Spending." Testimony prepared for the Subcommittee on Oversight of the House Permanent Select Committee on Intelligence, September 1980.

———. "Why the CIA's Estimate of Soviet Defense Procurement Was Off by 200 Percent: The Economic Consequences of Quality Change." First draft. August 1977.

References

Rostow, Eugene V. "Foreword." *Understanding the Soviet Military Threat*, pp. 1–6. Agenda Paper No. 6. National Strategy Information Center, Inc., 1977.

Rowen, Henry S. "Biting the Bullet and Other Hard Choices for Moscow." Draft paper. Revision of September 20, 1984.

———. "Prepared Statement of Hon. Henry Rowen." *Allocation of Resources in the Soviet Union and China, 1982: Hearings before the Subcommittee on International Trade, Finance, and Security Economics of the Joint Economic Committee of the United States*, pt. 8, exec. sess., June 29 and December 1, 1982.

Selin, Ivan. "Review of Soviet Military-Economic Analysis." Memorandum for the Director of Central Intelligence, June 22, 1981.

Shabanov, V. S. "Noviye podkhody (New Approach)." *Krasnaya zvezda*, August 18, 1989, pp. 1–2.

———. "Uvazhemiy Edward Sergeyivich (Dear Eduard Sergeyivich)." *Voprosy ekonomiki*, no. 10 (1990): 158–59.

Shevardnadze, Eduard. *The Future Belongs to Freedom*. Translated by Catherine A. Fitzpatrick. New York: The Free Press, 1991.

Shukhgal'ter, M. L. *Machino-stroitel'ny kompleks Rossii: trudny perekhod k rynko (The Machine-Building Complex of Russia: Difficult Transition to the Market)*. n.d.

*Soviet Defense Trends: Staff Study Prepared for Subcommittee on International Trade, Finance, and Security Economics of the Joint Economic Committee*. U.S. Congress, September 1983.

SSCI Professional Staff. "Soviet Defense Spending," Memorandum to DIA, October 15, 1987.

Steinberg, Dmitri. "Estimating Total Soviet Military Expenditures: An Alternative Approach Based on Reconstructed Soviet National Accounts." In Carl J. Jacobsen, ed., *The Soviet Defense Enigma*, pp. 27–57. Stockholm International Peace Research Institute: Oxford University Press, 1987.

———. "The Soviet Defense Burden: Estimating Hidden Defense Costs." *Soviet Studies* 44, 2 (1992): 237–64.

———. *Soviet National Accounting Methods and Practices*, IDS Report N2, September 1, 1988.

Steiner, James E. "CIA Estimates of Soviet Military Spending." *International Security* 14, 4 (Spring 1990): 185–93.

———. "Defense Economic Aspects of the Millar Report." Memorandum, January 28, 1992.

Steury, Donald, ed. *Sherman Kent and the Board of National Estimates: Collected Essays*. History Staff, Center for the Study of Intelligence, Central Intelligence Agency, Washington, D.C., 1994.

Stockholm International Peace Research Institute, *World Armaments and Disarmament: SIPRI Yearbook*. 1973–77, 1979–83, 1985–87, 1991.

Swain, D. Derk. "The Military Sector: How It Is Defined and Measured." In Henry S. Rowen and Charles Wolf, Jr., eds., *The Impoverished Superpower*, pp. 93–109. San Francisco: ICS Press, 1990.

Trimpin, William. "Estimated Cost of Soviet Deep Underground Construction." Memorandum for NIO/Strategic Programs, July 1, 1988.
U.S. Congress Joint Economic Committee. Subcommittee on International Trade, Finance and Security Economics of the Joint Economic Committee. *Allocation of Resources in the Soviet Union and China, 1983: Hearing,* 98th Cong., 1st sess., pt. 9, exec. sess., June 29 and September 20, 1983.
U.S. Congress Joint Economic Committee. Subcommittee on Priorities and Economy in Government. *Allocation of Resources in the Soviet Union and China: Hearing,* 93rd Cong., 2nd sess., exec. sess., April 12, 1974.
U.S. Department of Defense. *Soviet Military Power.* Washington, D.C.: U.S. Government Printing Office, Sept. 1981; Mar. 1983; Apr. 1984; Apr. 1985; Mar. 1986; Mar. 1987; Apr. 1988; Sept. 1989; Sept. 1990.
U.S. House of Representatives. *CIA Estimates of Soviet Defense Spending: Hearings before the Subcommittee on Oversight of the Permanent Select Committee on Intelligence.* U.S. Congress, September 3, 1980.
Wiles, Peter. "How Soviet Defense Expenditures Fit into National Income Accounts." In Carl G. Jacobsen, ed., *The Soviet Defense Enigma,* pp. 59–94. Oxford: Oxford University Press, 1987.
Wilson, George. "Intelligence Units Agree They Disagree." *Washington Post,* June 14, 1984, p. A10.
Wolf, Charles Jr. "The Costs and Benefits of the Soviet Empire." In Henry S. Rowen and Charles Wolf, Jr., eds., *The Future of the Soviet Empire,* pp. 121–40. New York: Institute for Contemporary Studies/St. Martin's Press, 1987.
———, Keith Crane, K. C. Yeh, Susan Anderson, and Ed Brunner. *The Costs and Benefits of the Soviet Empire.* Santa Monica, Calif.: RAND Corporation, 1986.
———, K. C. Yeh, Edward Brunner, Jr., Aaron Gurwitz, and Marilee Lawrence. *The Costs of the Soviet Empire.* Santa Monica, Calif.: RAND Corporation, 1983.
Working Group on Soviet Military Economic Analysis. *Report of the Methodology Panel of the Working Group on Soviet Military Economic Analysis.* July 1983.
Working Group on Soviet Military Economic Analysis. *Report of the Working Group on Soviet Military Economic Analysis.* July 20, 1983.
Zakharov, V. M. "Nuclear Defense within the System of Military War Prevention Measures." *Military Thought,* no. 2 (February 1994): 6–11.

# Index

Note: Pages with figures are indicated by italics; tables are indicated by underline.

Akhromeyev, Sergey, 139, 185, 186, 258$n$ 108
Alexander, Arthur, 185, <u>187</u>
Alsop, Joseph, 64–65
Altobelli, Steven, 237$n$ 47
Amory, Robert, 36
Andropov, Yuri, <u>122</u>, 127, 243$n$ 129
Arbatov, Georgiy, <u>122</u>, 259$n$ 115
Arms Control and Disarmament Agency (ACDA). *See* arms control and military-economic analysis: CIA support to ACDA
arms control and military-economic analysis: analysis of mutual constraints on defense budgets, 40, 41; CIA support to ACDA, 40; CIA support to Strategic Arms Limitation Talks (SALT), 40–41
Arms Control Verification Working Group. *See* arms control and military-economic analysis: CIA support to Strategic Arms Limitation Talks (SALT)
Aspin, Les, 39, 231$n$ 11
Atkeson, Edward, 77

Badgett, Lee, 232$n$ 20
Bakharov, G. A., 259$n$ 115
Bakhlanov, O. D., 258$n$ 103
Barry, James, 240$n$ 70
Becker, Abraham, 70, 140, 178, 182, 232$n$ 20, 236$n$ 30, 246$n$ 27, 250$n$ 26, 261$n$ 2
Becker panel. *See* Working Group on Soviet Military Economic Analysis (Selin panel/Becker panel)
Bennett, Donald, 46
Berenson, Paul, 145, 237$n$ 47
Birman, Igor, 143, 147, 157, 190, 252$n$ 47
Bissel, Schuyler, 255$n$ 81
Bobrovskiy, Sergey, 188
Bogomolov, Oleg, 259$n$ 118
Bond, Daniel, 183
Boretsky, Michael, 182–83, 257$n$ 96
Brandt, Willy, <u>122</u>
Brezhnev, Leonid, <u>122</u>, 126–27, 148
Brown, Harold, 145, 249$n$ 12
Brown, Nicholas, 46
building-block methodology: as alternative to DoD estimates, 26–27, 229$nn$ 23, 26; as alternative to official defense budget, 12, 228$n$ 7; and compatibility with CIA's economic intelligence on the

Soviet Union, 12–13, 153–55; computational requirements, 27–28, 48–49; computer model of, 42, 48–51, 52, 72, 73, 74–75, 97, 232*n* 24; and criticisms of CIA's definition of "national security" outlays, 146–48; defining "national security" outlays, 14, 146; discipline enforced on estimation process, 16, 50, 227*n* 3; easy to refine, improve, 16–18, 50; nature of, 11–12, 13–14; permits aggregation according to consumer requirements, 15–16, 228*n* 10; permits aggregation of disparate physical quantities, 3–4, 227*n* 1; and publications describing methodology, 33, 68, 236*n* 28; stages in development of, 30–33, *58*, 96–97, 209–22. *See also* CIA estimates of Soviet defense spending: charter to do; price estimates in building-block methodology; quantity estimates in building-block methodology; CIA estimates of Soviet defense spending, reliability of

burden. *See* defense burden, Soviet

Burton, Donald, 61, 73, 231*n* 11, 234*nn* 4, 6, 237*n* 47, 261*n* 6

Bush, George, 64, 218, 235*n* 12

*Business Week*, 64

Campbell, Robert, 232*n* 20

Carlucci, Frank, 189

Carter, Jimmy, 87, 115, 127, 201

Casey, William, 69, 74, 77, 88, 89, 94, 132

Chernenko, Konstantin, <u>122</u>

Chu, David, 65, 141, 235*n* 17

CIA estimates of Soviet defense spending: assessments of uncertainty of 1982 ruble price base, 98–99, 173, 174, 245*n* 1; charter to do, 9–10, 228*n* 4; cost of, 202–203; impact on national intelligence estimates (NIE's), 28–32, 43–44; lessons learned, 204–207; origins of CIA assessments of Soviet military-industrial strength, 5–6, 12–13, 26; origins of CIA given the responsibility, 9–10, 27, 31, 38; origins of National Security Council (NSC) interest, 6, 41; origins of need for independent estimates, 10, 26–27; presented at NATO, 140–41; reception in Pentagon, 31, 43, 54–55, 119; spending by military mission based on 1982 ruble prices, 106–107, 108, 110–11, *112*, 245*n* 5; spending by resource category based on 1982 ruble prices, 2, 101, 102–105, *106*; total spending based on 1982 ruble prices, 100, <u>101</u>, 102, <u>106</u>. *See also* CIA estimates of Soviet military spending, reliability of; Soviet defense programs, size of: CIA's dollar estimates

CIA estimates of Soviet military spending, reliability of: authors' assessment of, 192–94; capture of technological innovation, 152–53, 159–60, <u>161</u>, 166–67; confusion over stocks and flows, 103; consistency with data exchanged under arms control agreements, 158–59, 192; consistency with historical benchmarks, 121, <u>122</u>, 125–28; consistency with Soviet statements, 188–90, *189*, 195–96; dollar equivalent cost of personnel, O&M, construction, and RDT&E, 162–64, 253*n* 65, 254*nn* 66, 68; dollar equivalent costs of procurement, 159–60, 253*nn* 56, 57, 58, 60; findings of external review panels, 194–95; pre-1965 and post-1965, 98–99, 193; quantities, 158–59, 252*nn* 47, 48, 253*n* 55; ruble prices for procurement, 165–66, 254*n* 75; tests of, 168–73, <u>171</u>, <u>174</u>, 194, 254*n* 73

civil defense, Soviet, 218

Index

Clarke, Bruce, 42, 46, 89, 205–206, 232*n* 18
Cohn, Stanley, 183, 257*n* 96
Colby, William, 39
Columbia University, 131
construction: CIA estimates of Soviet outlays, 103, 163; evolution of estimates, 217–18, 263*nn* 17, 21, 22, 26
cost-effectiveness analysis, McNamara Pentagon, 35–36, 48
cost of empire. *See* defense burden, Soviet, expanded definition of
costing contracts with U.S. defense firms, 66, <u>67</u>, 97, 199–200, 213
criticisms of defense spending estimates (general), 8, 31, 45–46, 47; by CIA managers, 88–89, 119, 121; in measures of defense burden, 54–55; in NIE's, 46, 230*n* 37; outside CIA, 87; use of learning curves, 155. *See also* CIA estimates of Soviet military spending, reliability of; revisions in estimates, 1976, controversy and consequences of; Zumwalt, Elmo

Dam, Kenneth, 232*n* 20
Davies, R. W., 260*n* 126
defense burden, Soviet: analysis of constraints on defense spending and economic growth, 30, 85–86, 128, 135–39, 197, 247*n* 32, 248n, 40, 262*n* 13; basic estimate of, 129–31, <u>129</u>; expanded definition of, 132, <u>133</u>, *134*, 148, 243*n* 117, 247*nn* 28, 29, 249*n* 16; concepts and definition of, 21–23, 92–93, 130, 135–36, 245*n* 3, 246*n* 27, 248*nn* 35, 36; controversy over size of, 54–55, 72, 132, 233*n* 30, 247*n* 29; debate over influence of, on Soviet resource allocation, 86, 93–96, 137–39, 240*n* 82, 241*n* 97, 243*n* 129, 244*nn* 131, 133, 136, 246*nn* 18, 19; and dynamic measures, 135–36, 248*n* 33; and effect of distortions in Soviet prices, 81, 131–32, 142, 167, <u>168</u>, 190–91, 196; input-output analysis of, 55; overall assessment of CIA's measures of, 196–97; and perceptions of Soviet leadership, 142, 148, 188–90, 197, 244*n* 133, 260*n* 119; and Soviet incentives in arms control negotiations, 40, 126–28; and U.S. policy interest, 6, 9, 31, 92
defense industry, Soviet, 184–85, 258*nn* 103, 104, 260*n* 123
Defense Intelligence Agency (DIA): analysis of Soviet defense spending, 12, 37, 176–77; CIA cooperation with, 37–38, 39, 42, 66, 90, 158, 213; CIA differences with, 46, 89–90, 242*nn* 107, 109
Dementev, Valery, 259*n* 118
detente, 126–27
Duchene, Gerard, 255*n* 85
Dunn, Terrence, 238*n* 51

East, John, 239*n* 61
Eidelman, M. R., <u>184</u>, 258*n* 102
Eisenhower, Dwight, 229*n* 23
Enthoven, Alain, 36
Ericson, Richard, 232*n* 20
external assessments of CIA's military-economic work. *See* House Permanent Select Committee on Intelligence (HPSCI): panel report of; Military-Economic Advisory Panel (MEAP); President's Foreign Intelligence Advisory Board (PFIAB); Rowen Panel (Soviet Working Group); Working Group on Soviet Military Economic Analysis (Selin panel/Becker panel)

Fabian, Felix, 145
factor cost adjustments, 131, 142, 196
Firth, Noel, xiv, 73, 198, 234*n* 9, 249*n* 13

Fischer, Ben, 226*n* 22
Ford, Gerald, 39, 87, 122
Freeman, Jeff, 179

Garthoff, Raymond, 79
Gareyev, Makhmet, 251*n* 31, 260*n* 120
Gates, Robert, 40, 70, 72, 78, 86, 88–96, 119, 121, 148, 195, 213, 237*n* 33, 238*n* 53, 240*n* 82, 241*nn* 97, 98, 242*nn* 100, 106, 243*n* 129, 247*n* 29
Gaydar, Yegor, 238*n* 115
Gelman, Harry, 246*n* 15
Gershwin, Lawrence, 95
Gierek, Edward, 122
Godaire, John, 30, 38, 198, 230*n* 35
Goodpaster, Andrew, 229*n* 23
Gorbachev, Mikhail, 74, 84, 86–87, 95–96, 103, 104, 121, 122, 126, 128, 139, 175, 185, 188, 189, 195, 227*n* 4, 243*n* 129, 244*nn* 131, 133, 255*n* 77
Gostev, Boris, 259*n* 111
Graham, Daniel, 46, 55, 64–65, 228*n* 11, 255*n* 81
ground forces, CIA estimate of Soviet outlays, 107, 215
Greenslade, Rush, 136, 248*n* 35
Gushvin, Pavel, 188

Hanson, Philip, 65, 235*n* 21
Harvard University, 131
Hatch, Orin, 79
Heleniak, Timothy, 241*n* 98
Helgerson, John, 96–97, 145, 241*n* 98, 244*n* 140
Helms, Jesse, 239*n* 61
Helms, Richard, 46–47
Hewett, Ed, 232*n* 20
Hitch, Charles, 35–36
Holloway, David, 68, 126
Holzman, Franklyn, 65, 88, 91, 140, 148–49, 154, 160, 235*n* 19
House Permanent Select Committee on Intelligence (HPSCI): panel of, 140, 195: panel report of, 140, 158, 195, 261*nn* 1, 2, 262*n* 12

Huffstutler, Rae, 73, 77, 80, 88
Hunter, Holland, 47

Ignatius, David, 160–62
Ikle, Fred, 78, 90
index number problem, 148–50, 223–26. *See also* military programs, Soviet and U.S., comparison of size of: comparisons in rubles; Soviet defense programs, size of: CIA's dollar estimates
Intelligence Advisory Committee (IAC) Ad Hoc Military Cost Committee, 30–31
Intelligence Assumptions for Planning (IAP), 36, 231*n* 6

Jackson-Vanik amendment, 122
Jacobsen, Carl, 88, 162
Jepsen, Roger, 239*n* 61
Johnson, Lyndon, 122
Joint Analysis Group (JAG), 36
Joint Military Costing Review Board, 66, 91

Kaplan, Fred, 34, 200–201
Kaufman, Richard, 39, 242*n* 110
Kaufman, William, 87
Kennedy, John, 34, 125
Kent, Sherman, 30–31, 36
Kerr, Richard, 93
Khanin, Grigoriy, 184, 189–90, 258*n* 102
Khrushchev, Nikita, 103, 121–26, 122, 247*n* 32
Kirillin, V. A., 264*n* 27
Kissinger, Henry, 122
Kolt, George, 77–78
Komer, Robert, 30–31
Komorov, V., 258*n* 103
Kosygin, Aleksey, 122, 126
Kulikov, Viktor, 251*n* 31, 260*n* 120
Kushnirsky, Fyodor, 182
Kuznetsov, Vadim, 190

Lapidus, Gail, 232*n* 20

learning curves. *See* price estimates in building-block methodology: constant resource vs constant output prices
Lee, William, 65, 140, 143, 146, 157, 159–60, 175–76, 179–85, 195, 218, 238*n* 55, 249*n* 16, 253*nn* 55, 56, 57, 254*nn* 66, 68, 73, 255*nn* 79, 82, 256*nn* 87, 88, 257*n* 96, 258*n* 104, 261*n* 127, 262*n* 11, 263*n* 26
Levine, Herbert, 183, 232*n* 20
Ligachev, Yegor, 188–89, 195
Lobov, V. N., 190
Locher, James, 65, 87, 242*n* 103
Lopatin, V., 259*n* 115

MccGwire, Michael, 128, 246*n* 23
McCone, John, 38
McCurdy, David, 261*n* 2
MacEachin, Douglas, 86, 92, 119, 210, 244*nn* 131, 133, 134, 136
MacFarlane, Robert, 92
McMahon, John, 90
McNamara, Robert, 35, 37, 43, 48, 201
Mao Tse-tung, 122
Maresca, John, 119
Marshall, A. W., 92, 96–97, 145, 147, 160, 162, 199–200, 201, 233*n* 30, 237*n* 47, 244*nn* 139, 140, 247*n* 28, 253*n* 65
Martin, J. Michael, 237*n* 47, 238*n* 50
Medvedev, Zhores, 263*n* 22
Menshikov, Stanislav, 261*n* 127
Meyer, Stephen, 103, 232*n* 20
Michaud, Norbert, 147, 155, 234*n* 9, 242*n* 109
military analysis in CIA: origins of, 26–27, 41; impact on policy, 205; interactions with spending analysis, 3–4, 26–28, 70, 197–200; resources allocated to, 27–28, 42. *See also* utility of CIA estimates of Soviet defense spending: and focus on neglected elements of Soviet programs

Military-Economic Advisory Panel (MEAP): establishment of, 46–47; initial report of, 47, 54; subsequent recommendations by, 69–70, 80, 150, 168, 194–95, 236*n* 26, 237*n* 38. *See also* Working Group on Soviet Military Economic Analysis (Selin panel/Becker panel)
military programs, Soviet and U.S., comparisons of size of: CIA dollar estimates, 55, 112; CIA dollar estimates for 1951–64, 112–13, *113*, 115, 245*n* 6; CIA dollar estimates for 1965–89, 115–16, *116*, *118*, *129*, 245*n* 7; comparisons in rubles, 148–50, *149*, 243*n* 117, 250*n* 26; growing unpopularity, 87–93, 241*nn* 91, 98, 242*nn* 109, 110, 243*n* 111, 244*n* 140; use in debate over U.S. defense budget, xiii, 6, 35–36, 77–78, 87, 203–204. *See also* NATO-Warsaw Pact comparisons
Millar, James, 195
Millar panel. *See* House Permanent Select Committee on Intelligence (HPSCI): panel of
Millikan, Max, 12–13
mission, spending by: initial estimate of, 33; U.S. policy interest in, 35–36. *See also* CIA estimates of Soviet defense spending: spending by military mission based on 1982 ruble prices
misuse of CIA estimates, 4–5, 39–40, 203–204
Moiseyev, Michael, 141, 259*n* 111

Nagy, Dennis, 210
National Intelligence Council (NIC), 72, 77, 94, 238*nn* 52, 54
National Intelligence Officer for Strategic Programs (NIO/SP), 86, 95
National Security Agency (NSA), 66
NATO, 140–41

NATO-Warsaw Pact comparisons:
CIA estimates, 93, 116–19, 119, 120;
reception of, 119–20
navy, CIA estimates of Soviet outlays,
45–46, 110, 215–16
Nimitz, Nancy, 66
Niskanen, William, 232*n* 20
Nixon, Richard, 65, 87, 122, 126
Noren, James, xiii
Novoselov, Igor, 188

Odom, William, 147–48
Office of Computer Services (OCS),
49
Office of Data Processing (ODP), 74
Office of Economic Research (OER):
general, 42; contribution to
defense costing, 178; coordination
of work with Office of Strategic
Research (OSR), 42–43
Office of National Estimates (ONE),
26, 30–31, 36, 230*n* 43
Office of Research and Estimates
(ORE), 10
Office of Research and Reports
(ORR), 10, 12–13, 24, 26, 31, 41
Office of Soviet Analysis (SOVA), 71–
72, 73, 77, 78, 80, 84–86, 90, 91–96,
236*n* 31, 237*n* 38, 240*n* 82, 241*n* 91,
242*n* 107, 244*nn* 131, 134, 136
Office of Strategic Research (OSR),
42, 68, 73
Ogarkov, Nikolay, 122
Olmer, Lionel, 72
operations and maintenance, (O&M):
CIA estimates of Soviet outlays,
104, 162–63; evolution of estimates,
213, 214, 215–17, 263*nn* 11, 13, 16
organizations compiling Soviet
defense spending estimates in CIA:
Cost Analysis Branch, 42; Defense
Economic Division, 73; Defense
and Economic Issues Group, 73–
74; Defense Industries Division, 73;
Defense Management Branch, 74;

Military-Economic Analysis
Center, 73; Military-Economic
Planning Branch, 42; Military-
Economic Research Area, 41;
Military-Economics Branch, 41;
Military Expenditures Branch, 41.
*See also* Office of Research and
Reports (ORR); Office of Soviet
Analysis (SOVA); Office of
Strategic Research (OSR)
Ozhegov, A., 258*n* 103

Parker, Patrick, 147, 238*n* 55
Pavlov, Valentin, 259*n* 110
Percy, Charles, 138
personnel: CIA estimates of Soviet
outlays, 104, 166; evolution of
estimates, 209–11, 211, 212, 262*n* 5
Plummer, James, 232*n* 20
Poindexter, John, 92
politicization of defense spending
estimates: charges of, 6, 39–40, 78–
79, 87–88, 191–193: definition of, 7;
evaluation of charges, 7, 203
Posvar, Wesley, 232*n* 20
Pozharov, A. I., 251*n* 31
President's Foreign Intelligence
Advisory Board (PFIAB): report
of, 72, 94, 237*n* 41
price estimates in building-block
methodology: accounting for
inflation, 24–25; constant resource
vs. constant output prices, 155–56,
*156*, 252*nn* 42, 43, 44, 45; costing
directly in rubles, 24; in current
rubles, 81–82, *82;* fixed output
prices vs. real production costs, 71,
155–56, *156;* naval procurement and
O&M, 61; problems in maintaining
ruble price base, 25, 36–37, 51, 52–53,
56, 66, 152, 164–65, 233*n* 34; role of
dollar-to-ruble conversion ratios,
23–25, 99, 163, 229*n* 21; and
transition to 1982 ruble prices, 80–
81, 99, *100*, 150–54, 151, 250*nn* 27, 28,

29, 252*nn* 32, 33; and 1955 ruble prices, 52; and transition to 1967 ruble prices, 52, 53–54, 53, 232*n* 25; use of analogs, 18–19, 159, 162, 215, 228*n* 11 *See* military programs, Soviet and U.S., comparisons of size of: CIA dollar estimates; military programs, Soviet and U.S., comparisons of size of: costing contracts with U.S. defense firms; CIA estimates of Soviet military spending, reliability of

Proctor, Edward, 37–39, 41–42, 46–47, 54–55, 62, 89, 232*n* 18

procurement: CIA estimates of Soviet outlays, 102–103, 159–62, 165–66, *166;* evolution of estimates, 212–13, 263*n* 8; late call on plateau in spending, 75–80, 76, 89–91, 94, 237*n* 37, 238*nn* 52, 55, 239*nn* 58, 61, 69, 242*nn* 107, 109

"procurement paradox," 160, 162

Project Lamp, 36

projections of military forces and spending: and CIA reaction to Project Lamp, 36; and Intelligence Assumptions for Planning (IAP), 36; new approach to, 84–85, *85,* 240*nn* 75, 78; overestimates ("ramp effect") of, 83–84, 94, 199, 261*n* 5; reaction to new approach to, 85–87, 240*n* 82, 244*n* 134; underestimates, ("tired arm effect") of, 82, 199, 239*n* 68, 240*n* 72. *See* Project Lamp

Proxmire, William, 39, 64, 91, 231*n* 10

quantity estimates in building-block methodology: and gaps in information, 25–26, 157–58, 252*nn* 47, 48; sources of, 25–26, 51, 158–59, 192–93; use of analogs for, 19–20

RAND Corporation, 24, 34, 35, 36, 66, 131, 134

Rayzberg, B., 184, 259*n* 115

Reagan, Ronald, 77, 79, 87, 88, 122, 127, 201

reasonableness of dollar estimates, 169. *See also* CIA estimates of Soviet military spending, reliability of: tests of

reproducibility of dollar estimates, 168. *See also* CIA estimates of Soviet military spending, reliability of: tests of

research, development, test and evaluation (RDT&E): and CIA estimates of Soviet outlays, 105, 106, 163–64, 193, 221–22; and early CIA estimates, 61, 66, 68; and evolution of estimates, 219, *220,* 221–22; and MEAP recommendations, 68, 71; new approach to, 219, 221–22

residual methodology: CIA use of, 12, 177–78, 179–80, 256*n* 90; DIA use of, 12, 179, 256*n* 89; fundamental deficiencies of, 183–85, 190, 228*n* 5, 257*n* 101; nature of, 11, 177, 179, *180;* residual of Bond and Levine, in industrial statistics, 183; residual of Boretsky, in industrial statistics, 182–83; residual of Cohn, in industrial statistics, 183, 257*n* 96; residual of Lee, in industrial statistics, 179–82, 256*n* 88; residuals, in industrial statistics, 179–82, *181, 182,* 256*n* 88; residuals in national income, 177–79, 255*nn* 83, 85; residuals in state budget, 177, 255*n* 82. *See also* Soviet machinery prices, inflation in

resource category, spending by: first objective of CIA's estimate, 17, 33. *See also* CIA estimates of Soviet defense spending: spending by resource category based on 1982 ruble prices

revisions in estimates, 1976: causes of,

revisions in estimates, 1976 (cont.) collection of new prices, 61; growing consumer demand, 6, 41–42; neglect of ruble price base, 25, 58; role of emigre, 61–62, <u>63</u>, 64, 234n; controversy and consequences of, 57–58, 62, 64–65, 89, 234nn 8, 9; extent of, 59–61, <u>60</u>, 234n 3, 235n 11

Reynolds, John, 88

robustness of dollar estimates, 169–70, <u>171</u>, 213. *See also* CIA estimates of Soviet defense spending, reliability of: tests of

Rosefielde, Steven, 62, 88, 140, 147, 155, 157, 159–60, 179, 238n 55, 252nn 42, 48, 253nn 57, 58, 254n 75, 256n 88

Rostow, Eugene, 65

Rotmistrov, P. A., 125–26

Rowen, Henry, 72, 94, 146, 147, 157. *See also* Rowen Panel (Soviet Working Group)

Rowen Panel (Soviet Working Group), 72

ruble-dollar ratios. *See* price estimates in building-block methodology: role of dollar-to-ruble conversion ratios

Ryzhkov, Nikolay, 255n 77, 258n 103

Ryzhov, Yuri, 189

SCAM (Strategic Cost Analysis Model). *See* building-block methodology: computer model of

Schelling, Thomas, 232n 20

Schlesinger, James, 54, 233n 30, 234n 7

Schultz, George, 95–96, 244n 134

Selin, Ivan, 65, 70, 89–90, 116–17, 140, 162, 232n 20, 243n 111

Selin panel. *See* Working Group on Soviet Military Economic Analysis (Selin panel/Becker panel)

Selyunin, V., 259n 115

Shabanov, V. S., 251n 31, 258n 103

Shebarshin, Leonid, 258n 107

Shevardnadze, Eduard, 188–89, 195

Shukhgal'ter, Maria, 190

Smith, K. Wayne, 232n 20

Sonnenfeldt, Helmut, 249n 4

Soviet defense budget, official: CIA assessment of, 185–88; comparison with CIA estimates, 185; content of, 10–11, 175, 185–86, 191, 255n 76, 258n 106, 260n 126; defects of, 10–11, 12, 33, 175, 227n 4; revision in 1989 of, 185–87, <u>186</u>, 255n 77, 261n 127; Soviet assessments of, 188–90, 258n 108, 259nn 111, 115, 118, 260nn 120, 121; Western reliance on, CIA in early 1950s, 11; Western reliance on, DIA, 176–77, 255n 81; Western reliance on, Lee, 175–76, 255n 79; Western reliance on, Stockholm International Peace Research Institute (SIPRI) and International Institute for Strategic Studies (IISS), 175, *176*

Soviet defense programs, size of: CIA's dollar estimates, 4, 21, 22, 193–94, 223–26; conceptual meaning of, 21–22, 142–45, 229n 18; consistency in, 169; U.S. policy interest in, 6–7, 21, 31–32, 145–46. *See also* CIA estimates of Soviet military spending, reliability of: tests of

Soviet machinery prices, inflation in, 182–84, 257n 101, 258n 102

Soviet military prices, inflation in, 152–54, 166, 182, 250n 29, 251nn 30, 31, 254n 72

Soviet simulation (SOVSIM), 135–36

Steinberg, Dmitri, 191, 260n 124

Steiner, James, 238n 55, 261n 1, 262n 12

Strategic Arms Limitation Talks (SALT). *See* arms control and military-economic analysis: CIA support to Strategic Arms Limitation Talks (SALT)

Strategic Cost Analysis Model (SCAM). *See* building-block methodology: computer model of
Strategic defense: ABM deployment issue, 198; CIA estimates of Soviet outlays in rubles, 107
Strategic offense: CIA estimates of Soviet outlays in rubles, 107
Swain, D. Derk, 155, 238$n$ 55, 252$n$ 44
Symms, Steve, 239$n$ 61

tactical air: CIA estimates of Soviet outlays in rubles, 110, 216
Thurmond, Strom, 239$n$ 61
Treml, Vladimir, 232$n$ 20, 259$n$ 118
Turner, Stansfield, 69

U.S. Congress: CIA testimony before Joint Economic Committee, 38–40, 90–91, 92, 93, 94. *See also* Aspin, Les
utility of CIA estimates of Soviet defense spending: blueprint for future analysis, 204; contribution to historical analysis, 202; in cost-benefit analyses, 6, 197–98; in debate over U.S. defense budget, 6, 87–88, 91, 200–202; and focus on neglected elements of Soviet military programs, 18, 199–200, 202, 217, 249$nn$ 12, 13; general, 3–4, 6, 45, 141–42, 200; intelligence on Soviet design and manufacturing practices, 199–200; Military-Economic Advisory Panel (MEAP) appraisals, 47, 69, 141, 236$n$ 30; projecting forces and deployment, 197–200; for U.S. government departments and agencies, 91–92, 141, 145–46, 200–202, 249$n$ 4; views of critics, 142–43, 145. *See also* arms control and military-economic analysis

Vance, Cyrus, 38
Vance-McCone accord, 38
Voronin, L. A., 258$n$ 103

war scare in 1981–82, 122, 127, 246$nn$ 21, 22
Weinberger, Casper, 77–78, 88, 89, 91–92, 96, 247$n$ 29
Weiss, Stanley, 232$n$ 20
Wiles, Peter, 255$n$ 85
Working Group on Soviet Military Economic Analysis (Selin panel/Becker panel), 70, 73, 89, 140, 147, 155, 157, 160, 162, 237$n$ 33; Report of the Methodology Panel of the Working Group on Soviet Military Economic Analysis, 71–72, 160, 165; Report of the Working Group on Soviet Military Economic Analysis, 70–71, 89–90, 157–58, 183, 195, 237$n$ 37, 242$nn$ 99, 100

Yazov, Dmitri, 189

Zakharov, V. M., 259$n$ 115
Zhuravlev, Anatoliy, 260$n$ 121
Zumwalt, Elmo, 45–46